Rishi Paliwal • Shivani Rai Paliwal

Advances in Nanochemoprevention

Controlled Delivery of Phytochemical Bioactives

Springer

Rishi Paliwal
Nanomedicine and Bioengineering
Research Laboratory, Department
of Pharmacy
Indira Gandhi National Tribal University
Amarkantak, Madhya Pradesh, India

Shivani Rai Paliwal
SLT Institute of Pharmaceutical Sciences
Guru Ghasidas Vishwavidhyalaya
(A Central University)
Bilaspur, Chhattisgarh, India

ISBN 978-981-15-9691-9 ISBN 978-981-15-9692-6 (eBook)
https://doi.org/10.1007/978-981-15-9692-6

This Springer imprint is published by the registered company Springer Nature Singapore Pte Ltd.
The registered company address is: 152 Beach Road, #21-01/04 Gateway East, Singapore 189721, Singapore

To our mentor "Prof. Suresh Prasad Vyas," our parents "Mr. Navin Chandra Paliwal-Mrs. Chanchla Paliwal and "Mr. Laxmi Prasad Rai-Mrs. Anita Rai," and our lovely sons "Aadi" and "Meeku"

Preface

This book comprises recent advances in the field of nanochemoprevention focusing on Controlled Delivery of Phytochemical Bioactives. Chemoprevention, utilizing natural dietary compounds, has regained interest due to a larger safety window and proven efficacy of such molecules in cancer treatments. Many compounds of natural origin, which possess excellent anti-oxidant and anti-inflammatory activity, have shown chemo-preventive potential and remained capable to synergize the therapeutic effect of cancer chemotherapy alone as well as upon co-administration in combinations. Utilization of chemo-preventive potential of such compounds is limited due to their limited water solubility resulting in low bioavailability and poor therapeutic efficacy. Nanotechnology has revolutionized drug delivery especially in case of dreadful diseases like cancers through passive and active targeting. Nanoparticles are well-known tiny drug carrier/cargos known for their small size and unique capability to deliver numerous bioactive agents in the vicinity of diseased cells or tissues. Due to high payload, surface functionalization opportunity, improved biodistribution, and minimum side-effects to healthy cells, they are the preferred choice to fight against cancer.

Recently, higher attention has been paid by scientific fraternity to prepare biomaterial based nanoparticles bearing such drugs/agents to achieve their higher chemo-preventive action both ex-vivo and in vivo through controlled delivery. Such an expanding arena of nanochemoprevention research needs timely updates and review to concise research outcomes. This book summarizes chapters on classification of natural molecules, type of nanoparticles tested and their controlled release, mechanistic aspects of their superior efficacy over plain drug molecules, and the progress and updates of pre-clinical results of developed formulations for cancer chemoprevention. A chapter is exclusively dedicated to discussion of targeted nanomedicine of phytoconstituents. Further, scale-up and regulatory aspects of the development of nanoparticles and herbal products are also highlighted in a separate chapter to understand the commercialization capabilities of nanochemopreventive products developed so far.

The book provides comprehensive insight for early researchers working in the area of nanochemoprevention. This book will be an explicit asset for the undergraduate, master students, and research fraternity working into areas of herbal product, biomaterials sciences, and oncology particularly for plant based product development of cancer prevention or for the value addition of the phytoconstituents using nanotechnology towards achieving their superior chemopreventive activity.

We are thankful to our respective institutes Indira Gandhi National Tribal University, Amarkantak and Guru Ghasidas Vishwavidyalaya (A Central University), Bilaspur for providing necessary resources and support to write this book. The idea and execution were inspired by natural resources and biodiversity of the Holy Amarkantak region, which are generously available in the nearby surroundings of our institutes. The support of Mr. Rameshroo Kenwat is duly acknowledged. We are thankful to Dr. Bhavik Sawhney from Springer Nature for his continuous follow-up with full of patience.

Amarkantak, Madhya Pradesh, India Rishi Paliwal
Bilaspur, Chhattisgarh, India Shivani Rai Paliwal

Contents

Chemoprevention: A General Introduction 1

Abstract

This chapter describes the basic introduction of carcinogenesis and chemoprevention. A section presented highlights on the natural resources of chemopreventive agents. The different mechanism pathways of chemoprevention are also included in this chapter. Pharmaceutical need for the development of safe and more efficacious chemoprevention strategies is also discussed. A brief account on nanotechnology-based delivery of chemopreventive agents is discussed.

Keywords

Cancer prevention · Chemotherapy · Natural resources · Herbal anticancer drug · Nanotechnology

1.1 Introduction

Cancer is one of the leading cause of deaths worldwide. Cancer statistics shows that a large number of patients are coming up every year with an increasing number day by day. As per World Health Organization more than 14 million people suffer from cancer and about eight million die worldwide. Several treatment options are in clinical practice; among them chemotherapy is included in almost every treatment regime. Unwanted side effect of cytotoxic agent due to non-target organ distribution of the drug affects patient compliance of chemotherapy (Baldo and Pham 2013). Nanomedicine based on novel drug delivery system has improved drug pharmacokinetics and direct it towards site-specific drug delivery to the cancerous cells. However, it increases the cost of the overall treatment and also remains less effective in case of cancer metastasis.

R. Paliwal, S. R. Paliwal, *Advances in Nanochemoprevention*,
https://doi.org/10.1007/978-981-15-9692-6_1

Chemoprevention has attracted attention of scientific community in order to search natural phytoconstituents as alternative to synthetic drugs for cancer management (Wang et al. 2012). Since natural plant based constituent serves as a safer and cost-effective option, it has been popular and well accepted by the patients. Further, in this case here focus is given more on cancer prevention than treatment, hence patient compliance is high. Natural phytochemicals not only prevent, treat, delay but also provide care to the cancer patients in many ways. Several mechanisms have been reported that how these phytoconstituents are active in chemoprevention (Kaur et al. 2018). Plants having chemopreventive agents are part of our dietary consumption. As an alternative medicine, they have been advised since ancient time. Many of the phytoconstituents are available in the market as dietary supplement. Such chemopreventive dietary supplements are being advised to cancer patients worldwide as adjunct therapy (Norman et al. 2003).

1.2 Carcinogenesis

The term "carcinogenesis" is defined as a multi-step biological event that takes place due to occurrence of genetic events starting from a single cell leading to uncontrolled cell growth. "Chemoprevention" is a widely used expression which defines the inhibition of either or both cancer initiation, development, and/or progression using chemicals that function on various stages of carcinogenesis (Fitzpatrick 2001). Such chemopreventive molecules may be naturals; synthetic or biologic chemical agents that block the DNA damage or arrest/revert the progression of premalignant cells. In other words, chemoprevention is the approach purposely designed to prevent or reverse the process of carcinogenesis before it starts invasion and metastasis using pharmacological agents (Sporn and Suh 2002). Usually, normal cell converts into a neoplastic cell upon exposure to carcinogenic and mutagenic agents leading to tumorigenesis by promotion and progression. In promotion phase, initiated cells transform to preneoplastic cells by cancer promoters and progression involves conversion of preneoplastic cells to neoplastic cells (Sporn and Suh 2000; Greenwald 2002). Successively with time tumor mass may then metastasize to other biological locations passing through circulatory system.

1.3 Chemoprevention

Generally, "chemoprevention" involves the use of agents which reverse, suppress, or prevent the transformation of preneoplastic cells to neoplastic cells. Therefore, chemoprevention can be achieved mainly by two strategies: one by preventing carcinogen mediated cell initiation by blocking agents and second by impairing the neoplastic transformation by suppressing agents (Surh 2003). Chemoprevention can be achieved by understanding cell signaling and molecular pathways involved in the initiation and progression of cancer and interfering with them so that normal cell do not convert to neoplastic cell. Ideally, a chemopreventive agent must be efficacious, act on multiple molecular targets with known mechanism of action, economically acceptable, safe and non-toxic on long-term use. Mehta et al. 2010 reviewed on various studies for cancer chemoprevention using natural products. Authors classified chemopreventive agents on the basis of experimental carcinogenesis models in a stage-specific manner.

1.4 Natural Resources of Chemopreventive Agents

Nature serves as the resources for variety of materials required for food, textile, medicine, agriculture, etc. Since ancient time, a large number of populations depend upon plants either for treatment or prevention of diseases and ailments. The plant resources are cheap, safe, affordable, and accessible for long-term treatments as well. Dietary prevention of the cancer remains integral part of earlier food habits and is closely associated with our natural medicine system. Phytochemicals belonging to polyphenols, flavonoids, and their sub-classes are highly efficacious in chemoprevention due to their antioxidant, anti-inflammatory, and antiproliferative biological activities. Most commonly tested compounds known for chemopreventive activity are curcumin, epigallocatechin-3-gallate, genistein, resveratrol, thymoquinone, luteolin, caffeic acid, wogonin, etc. Chemical structures of some of these compounds are presented in Fig. 1.1. These phytochemicals hinder the growth of cancer by interfering number of signaling pathways and molecular sites and can be considered as promising therapeutic agents for prevention and treatment (Bode and Dong 2004). However, these phytochemicals and nutrients suffer with the problem of low bioavailability due to their hydrophobic nature and hence need higher dose to exert their therapeutic action that limits their wide applicability.

In a study, National Cancer Institute reported a vital link between role of food and its nutrients with cancer chemoprevention. It was reported that regular consumption of fruits, vegetable, and the whole grains may reduce risk of different types of cancers. For example, green tea, lycopene, modified citrus pectin, pomegranate, soya, VitD, VitE, Calcium, Selenium and their combinations are used for the prostate cancer patients. https://www.cancer.gov/about-cancer/treatment/cam/patient/prostate-supplements-pdq).

Zhang and co-workers reported that traditional Chinese medicines based on plants may be used as complementary and alternative medicine for cancer and diabetes (Zhang et al. 2013). Some of the reported TCM plants are *Panax ginseng, Pinellia ternata, Salviae miltiorrhizae, and Arisaema japonicumare.* About 80 species of Polygonum are being used in TCM. Authors reported that polyphenols present in fruit and vegetables have demonstrated multiple effects in chronic diseases including cancers. Wang and co-workers summarized some more natural bioactive compounds that include genistein from soyabean, curcumin from turmeric, and resveratrol from grapes, epigallocatechin from green tea, sulforaphane from broccoli, silymarin I from milk thistle, phenethyl isothiocyanates from cruciferous vegetables, diallyl sulfide from garlic, lycopene from tomato, apigenin from parsley, rosmarinic acid from rosemary, Vitamin E from wheat germ oil and sunflower oil,

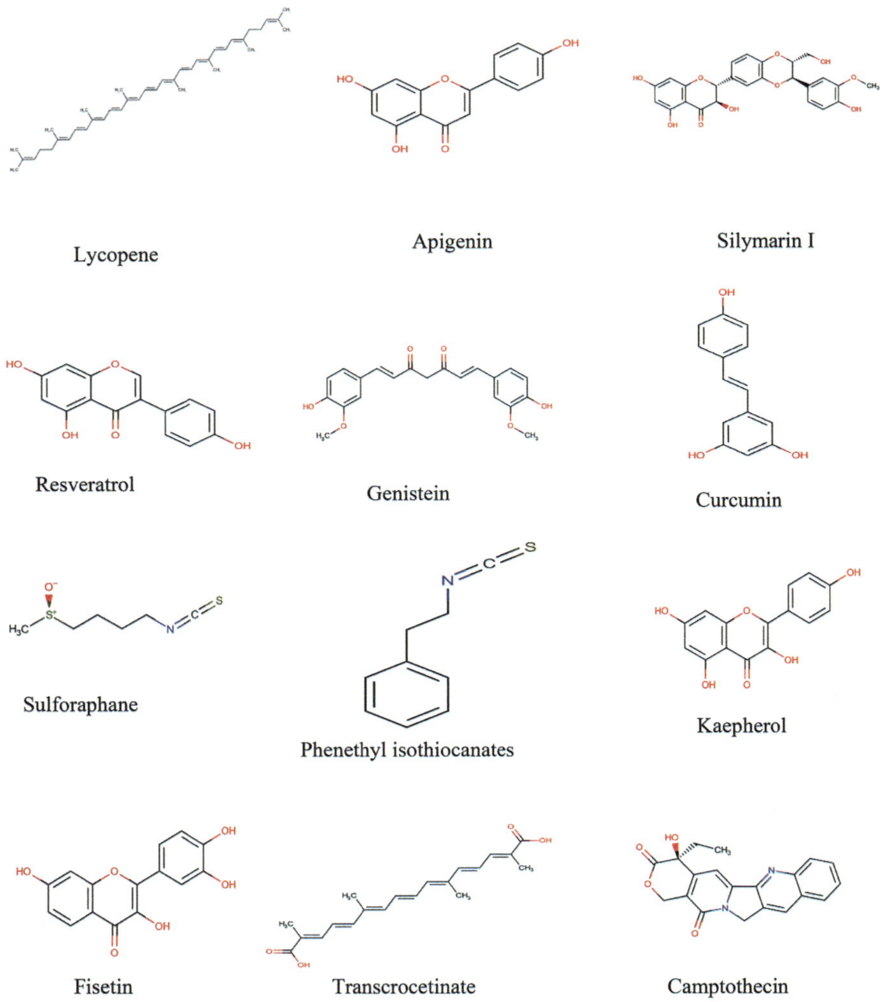

Fig. 1.1 Chemical structure of some known chemoprotective phytochemicals (Structures are adopted from drug bank database)

Vitamin D from mushroom, kaempferol from tea, broccoli, and grape fruit, fisetin from strawberries, apples, diindolylmethane.indole-3-carbinole from broccoli cauliflower, collard greens, crocetin from saffron, and gingerol from ginger (Wang et al. 2012). The mechanism of action of these phytochemicals with pharmaceutical challenges and opportunities has also been reported in the literature.

1.5 Mechanism Pathway for Chemoprotective Effect

Mechanistically, chemoprotective effect of phytochemicals is exerted via different pathways including antioxidative and anti-inflammatory activities along with apoptosis, authophagy, and cell cycle arrest in different phases and induction of phase II enzymes (Zhao et al. 2018). A large number of mechanisms of action have been described in the literature. More than 100 oncogenes have been identified till date. For example, Racz et al. 2002 reported chemoprotective activity of BGP-15 through modulation of poly (ADP-ribose) polymerase (PARP) activity leading to protection of mitochondria from oxidative damages (Racz et al. 2002). In another study, the role of NF-E2-related factor 2 (Nrf2) in chemoprevention was confirmed on Nrf2 deficient mice which were lacking in response to some chemopreventive agents (Hu et al. 2010). Similarly, few chemopreventive compounds have shown inhibition of polo-like kinase 1 (PlK1) in cancer cell (Schmit et al. 2010).

1.6 Nanotechnology for Delivery of Chemopreventive Agents

In order to search more specific drug delivery approaches for cancer treatment, nanotechnology provide solutions for most of the conventional therapy limitations. Controlled release, cellular targeting, and intracellular delivery of chemopreventive agents could be achieved using such engineered nanosystems. Tiny drug cargos avoid the normal cell and target only desired cancerous cells. Further, phytochemicals have poor aqueous solubility issues which render their suitable pharmaceutical formulation. Nanotechnology may be a useful tool to overcome such limitations of poor absorption of chemopreventive agents (Muqbil et al. 2011). Nanoparticles can be successfully utilized for augmented delivery of these natural products especially designed for cancer prevention and treatment (Mehta et al. 2010). Nanoparticles are tiny cargos well-known for increased circulation time of the loaded chemotherapeutic agent with increased efficacy for drug localization at target site and decreased multi-drug resistance (MDR) (Tekchandani et al. 2017). This is why an emerging focus on developing nanoparticles of chemopreventive agents has arisen among the scientific community. This review discussed about importance of such phytoconstituents and the advancement in the direction of exploring the role of various nanoparticles to potentiate the therapeutic output for chemoprevention. Figure 1.2 depicts various aspects related to prevention and cure of cancer and utilization of nanoparticles in controlled delivery strategies to combat tumor/cancer cells.

1.7 Conclusion

Chemoprevention is diminishing the progression of the cancer using chemopreventive agents. Antioxidants and anti-inflammatory agents of either natural or synthetic sources act via different mechanisms. Curcumin, resveratrol,

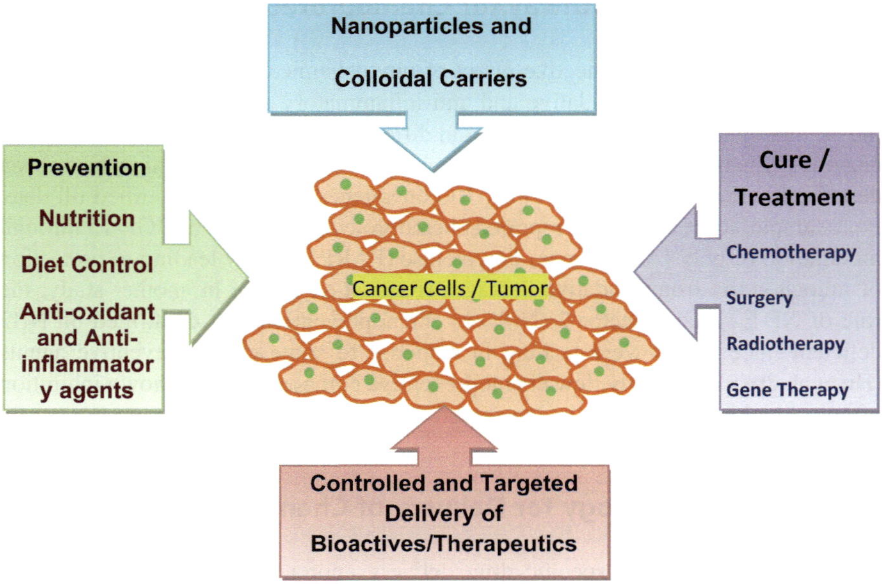

Fig. 1.2 Cancer chemoprevention, therapy options, and drug delivery options to combat its progression

lycopene, silymarin, vitamin A and E, and camptothecin are some natural chemopreventive molecules. However, most of these molecules suffer with the poor bioavailability and stability issues. Loading or encapsulation of these molecules into nanoparticles limits the problems associated with the clinical use of these natural phytochemicals. Using combination of both nanotechnology and herbal technology for cancer prevention and progress is called nanochemoprevention. It is evident from the literature that scientific fraternity is looking for more product oriented nanopharmaceuticals for chemoprevention.

References

Baldo BA, Pham NH (2013) Adverse reactions to targeted and non-targeted chemotherapeutic drugs with emphasis on hypersensitivity responses and the invasive metastatic switch. Cancer Metastasis Rev 32(3–4):723–761

Bode AM, Dong Z (2004) Targeting signal transduction pathways by chemopreventive agents. Mutat Res 555(1–2):33–51

Fitzpatrick FA (2001) Inflammation, carcinogenesis and cancer. Int Immunopharmacol 1 (9–10):1651–1667

Greenwald P (2002) Cancer chemoprevention. BMJ 324(7339):714–718

Hu R, Saw CLL, Yu R, Kong ANT (2010) Regulation of NF-E2-related factor 2 signaling for cancer chemoprevention: antioxidant coupled with antiinflammatory. Antioxid Redox Signal 13 (11):1679–1698

Kaur V, Kumar M, Kumar A, Kaur K, Dhillon VS, Kaur S (2018) Pharmacotherapeutic potential of phytochemicals: implications in cancer chemoprevention and future perspectives. Biomed Pharmacother 97:564–586

Mehta RG, Murillo G, Naithani R, Peng X (2010) Cancer chemoprevention by natural products: how far have we come? Pharm Res 27(6):950–961

Muqbil I, Masood A, Sarkar FH, Mohammad RM, Azmi AS (2011) Progress in nanotechnology based approaches to enhance the potential of chemopreventive agents. Cancers 3(1):428–445

Norman HA, Butrum RR, Feldman E, Heber D, Nixon D, Picciano MF, Rivlin R, Simopoulos A, Wargovich MJ, Weisburger EK, Zeisel SH (2003) The role of dietary supplements during cancer therapy. J Nutr 133(11):3794S–3799S

Racz I, Tory K, Gallyas F Jr, Berente Z, Osz E, Jaszlits L, Bernath S, Sumegi B, Rabloczky G, Literati-Nagy P (2002) BGP-15—a novel poly (ADP-ribose) polymerase inhibitor—protects against nephrotoxicity of cisplatin without compromising its antitumor activity. Biochem Pharmacol 63(6):1099–1111

Schmit TL, Ledesma MC, Ahmad N (2010) Modulating polo-like kinase 1 as a means for cancer chemoprevention. Pharm Res 27(6):989–998

Sporn MB, Suh N (2000) Chemoprevention of cancer. Carcinogenesis 21(3):525–530

Sporn MB, Suh N (2002) Chemoprevention: an essential approach to controlling cancer. Nat Rev Cancer 2(7):537

Surh YJ (2003) Cancer chemoprevention with dietary phytochemicals. Nat Rev Cancer 3(10):768

Tekchandani P, Kurmi BD, Paliwal SR (2017) Nanomedicine to deal with cancer cell biology in multi-drug resistance. Mini Rev Med Chem 17(18):1793–1810

Wang H, Oo Khor T, Shu L, Su ZY, Fuentes F, Lee JH, Tony Kong AN (2012) Plants vs. cancer: a review on natural phytochemicals in preventing and treating cancers and their druggability. Anti-Cancer Agents Med Chem 12(10):1281–1305

Zhang H, Li C, Kwok ST, Zhang QW, Chan SW (2013) A review of the pharmacological effects of the dried root of Polygonum cuspidatum (Hu Zhang) and its constituents. Evid Based Complement Alternat Med 2013:208349

Zhao Y, Hu X, Zuo X, Wang M (2018) Chemopreventive effects of some popular phytochemicals on human colon cancer: a review. Food Funct 9(9):4548–4568

Phytochemical Bioactives in Chemoprevention

2

Abstract

This chapter deals with the information on natural nutrients used for chemoprevention. Phytochemicals such as polyphenols (curcumin), flavonoids (anthocyanidins, flavonols, flavones, flavanones, catechins (Flavan-3-ols), and isoflavones), phytoalexin (resveratrol), and kaempferol are discussed in detail with their mechanisms as chemopreventive agents. Various signaling pathways or factors which are reported in the literature have been described. Various properties of nutrients and phytochemicals under investigation as chemopreventive agents have also been elaborated.

Keywords

Phytoconstituents · Anticancer plant · Anti-inflammatory action · Mechanism of action · Molecular targets

2.1 Natural Nutrients for Chemoprevention: An Introduction

Several natural or synthetic molecules are known to exert chemopreventive action via different mechanism of their action (Table 2.1). This is very interesting that since ancient time in our food habits, we are using plenty of natural products in one or another form and they possess certain nutrients that are beneficial for regulation and maintenance of healthy cells. It has been noted that an improper food culture may be associated with increased risk of cell proliferation resulting into higher risk of certain type of cancers (Lucenteforte et al. 2009). Natural compounds remain present in the form of secondary metabolites in the food materials like alkaloids, phenylpropanoids, flavonoids, isothiocyanates (ITC), and isoprenoids (Iriti and Faoro 2009a). Numerous pharmacological activities have been shown by these metabolites such as antioxidant, anti-inflammatory, pro/anti-apoptotic, and

R. Paliwal, S. R. Paliwal, *Advances in Nanochemoprevention*,
https://doi.org/10.1007/978-981-15-9692-6_2

Table 2.1 List of chemopreventive agents and their different mechanism of chemoprevention

Category	Agents	Example	Type of cancer	Reference
NSAIDS	Cyclooxygenase-2 inhibitors	Celecoxib, Parecoxib	Colorectal adneocarcinoma, transitional cell carcinoma of the bladder, breast adenocarcinoma, cervical intra-epithelial neoplasia, lung cancer, skin cancer and oral leukoplakia	Bresalier et al. (2005) and Nussmeier et al. (2005)
Molecularly targeted anti-carcinogenic mechanism	Peroxisome proliferation activator receptor gamma agonist	Rosiglitazone	Hormone receptor negative breast cancer	Mehta et al. (2000)
	EGFR inhibitors	EGFR antibodies	Colorectal head and neck cancer	Shin et al. (1994)
	P53 targeting	ONYX-015 adenovirus	Oral	Rudin et al. (2003)
	DNA methylation inhibitor	Hydralazine	Leukoplakia (P53 deficient cells) Oral leukoplakia	Deng et al. (2003)
	Histone deacetylase inhibitor	Synthesized hydroxamic acid	Liver cancer cells	Marks et al. (2000)
Phytochemicals	Polyphenols	Curcumin	Breast, ovary, prostate, pancreas and skin	Li et al. (2005)
	Phytoalexin	Resveratrol	Myeloma, breast, ovary	Brisdelli et al. (2009)
Combination therapy	EGFR signaling inhibitor and COX inhibitor	EKI-569 and Sulindac	Colonic neoplasia	Torrance et al. (2000)
	HMG CO-A reductase inhibitor and NSAIDS	Lovastatin and Sulindac	Preneoplastic cells to increase apoptosis by caspase 3 activity	Agarwal et al. (1999)
	Polyamine and NSAIDS	Difluoromethylornithine (DFMO) and Sulindac	Colorectal adenocarcinoma	Thompson et al. (2010)
	NSAIDS and phytochemicals	Aspirin, curcumin, and sulforaphane	Pancreatic cancer	Grandhi et al. (2013)
Other agents	Selenium	Selenium nanoparticles	Skin carcinoma	Clark et al. (1996)

vasodilating activities, which are directly or indirectly related to their chemopreventive action (Table 2.2) (Iriti and Faoro 2009b).

National Cancer Institute screened about 35,000 plant species for potential cytotoxic activities among them about 3000 plant species were having reproducible anticancer activity (Desai et al. 2008). Some of the bioactive from plants are 20 β-hydroxyecdysterone, cordioside, columbin from *Tinospora cordifolia*, betulin, betulinic acid from *Ziziphus nummularia*, andrographolide from *Andrographis paniculata*, asiaticoside, hydrocotyline, vallerine, pectic acid, sterol, stigmasterol, flavonoids, thankunosides and ascorbic acid from *Centella asiatica*, curcumin from *Curcuma longa*, nirtetralin, niranthrin, phyllanthin, phyltetralin from *Phyllanthus amarus*, bullatacin from *Annona atemoya*, camptothecin from *Mappia foetida*, withaferin A from *Withania somnifera*, lignans, wikstromol, matairesinol, and dibenzyl butyrolactol from *Cedrus deodara* and triterpenic acids from *Boswellia serrata* (Desai et al. 2008). Although most of these compounds have been evaluated for anticancer potential on cell culture studies and not on humans, their potential cannot be ignored without further studies. Recently, Komakech et al. (2017) reviewed potential chemo preventive and chemotherapeutic effect of plant derived phytochemicals from *Prunus africana* (Hook f.) bark. This plant is traditionally being used for treating prostate cancer in Africa. Authors reported both in vitro and in vivo antiprostate cancer effect of these phytochemicals. Mechanistically, phytochemicals present in plants extract were having strong antiandrogenic and antiangiogenic activities. The study suggested carrying out further pre-clinical and clinical studies to establish the promising potential of phytochemicals present in this plant extract for both prevention and chemotherapy in human prostate cancer. In another study, aqueous and alcoholic extracts of *Solanum nigrum* were tested pre-clinically on N-nitrosodiethylamine (NDEA)-induced hepatocellular carcinoma (HCC) rat model (Akshatha et al. 2018). Authors observed immunohistochemical and histopathological changes and chemoprotective effect on Wistar rat animals 28 days after oral administration of these plant extracts. Authors reported that alcoholic extract reduced the severity of lesions in the liver comparable to standard drug sorafenib. Further, authors claimed that the immunoreactivity of the hepatocytes treated with a higher dose of alcoholic extract of *S. nigrum* was limited and was comparable to a standard drug, sorafenib. Figure 2.1 shows the various signaling pathways that serve as target in cancer for phytochemicals.

2.2 Phytochemicals in Chemoprevention

Figure 2.2 demonstrates steps involved in the cancer initiation and progression, and role of phytochemicals where they exercise their action. Such phytochemicals have diverse molecular and biochemical targets in healthy and diseased conditions (Table 2.3). Even being such useful for chemopreventive action, these molecules compromise with poor potency and bioavailability. An approach of using combination of two or more phytochemicals together has been advocated as such or after loading them into a suitable drug delivery cargo. Due to required high doses for their

Table 2.2 Properties of nutrients and phytochemicals investigated for cancer chemoprevention

Phytochemicals or bioactives	Source	Physico-chemical class	Pharmacological activities	Mechanism of action in chemoprevention	Reference
Curcumin	*Curcuma longa*	Polyphenol	Chemopreventive, anti-inflammatory, anti-oxidative, and anti-carcinogenic activities	It suppresses NF-κB activation and NF-κB gene products and can induce p53-dependent apoptosis by induction of p53 in certain cancer cell lines	Li et al. (2005)
Epigallocatechin-3-gallate	Green tea is produced from the unfermented leaves of *Camellia sinensis*	Polyphenol	Antioxidant, anti-inflammatory, and anti-tumorigenic	p53-dependent and Fas-mediated pathways	Kuo and Lin (2003)
Genistein	Phytoestrogen of Genista tinctoria, Soybean	Isoflavones	Antiproliferative activity, cardiovascular disorders, menopausal symptoms, osteoporosis	Induction of apoptosis, reversal of epithelial mesenchymal transition	Zhang et al. (2008)
Resveratrol	Grapes, berries, peanuts, and red wine	Phytoalexin	Antioxidant, anti-aging, antiviral, cardiovascular and neuroprotective effects, antiproliferative, anticancer and chemopreventive	Different molecular switches involved in carcinogen metabolism (either activation and detoxification), inflammation, cell proliferation, cell cycle, apoptosis, angiogenesis, cardio protection, neural tissue degeneration, tumor metastasis	Brisdelli et al. (2009) and Xu et al. (2015)
Thymoquinone	Seed of *Nigella sativa*	Benzoquinones	Anti-inflammatory, antioxidant, and anti-carcinoma effects	Induction of apoptosis resulting from mitochondrial dysfunction	Salim et al. (2013)

Wogonin	Dry herb *Scutellaria baicalensis*	Flavonoids	Anti-inflammatory, antiviral activities, anticancer activity	Apoptosis through the mediation of Ca2+ and/or inhibition of NF-κB	Tan et al. (2011)
Luteolin	Green vegetables such as artichoke, broccoli, cabbage, celery, cauliflower, green pepper, and spinach	Hydroxyflavone	Anti-inflammation, anticancer effects	It triggers apoptotic cell death by activating apoptosis pathways and suppressing cell survival pathways	Majumdar et al. (2014)
Caffeic acid	Plants, fruits, vegetables, olive oil, and coffee	Organic compound belongs hydroxycinnamic acid class	Antioxidant, Immunomodulatory and anti-inflammatory activity, anticancer activity or carcinogen?	Increased oxidative DNA damage?	Prasad et al. (2011)
Sulforaphane	Cruciferous vegetables such as broccoli, Brussels sprouts, cauliflower, and cabbage	Sulfur-containing isothiocyanate	Chemopreventive, anti-oxidative and anti-carcinogenic activities	SFN has been shown to reduce NF-κB activity and affect expression of NF-κB mediated genes encoding adhesion molecules, inflammatory cytokines, growth factors and anti-apoptotic factors. SFN also modulates multiple targets, which regulate many cellular activities including oxidative stress, apoptosis induction, cell cycle arrest, angiogenesis, and metastasis suppression	Kallifatidis et al. (2009)

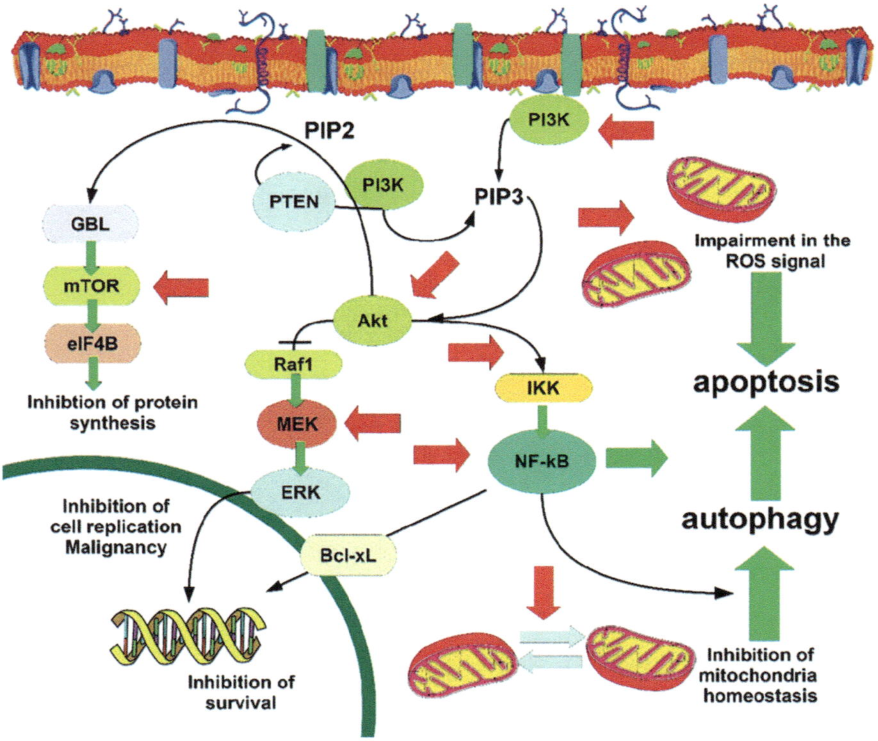

Fig. 2.1 Schematic representation of the major signaling pathways targeted by phytochemicals in cancer. Phytochemicals principally act as inhibitors which red arrows indicate in the PI3K/Akt signaling pathway. Green arrows indicate the action promoted or activated by flavonoids, also following the inhibitory signaling cascades (Adopted from Chirumbolo et al. 2018)

pharmacological action, a suitable strategic drug delivery carrier may be useful tool for better therapeutic action. Additionally, combination therapy remains always superior as it avoids drug resistance development after long-term use of single high dose phytochemical due to diverse mechanism of action of each molecule. Chemical structure of some well-known chemopreventive compounds has been shown in Fig. 2.3 and details of such molecules are described in the following sections.

2.2.1 Polyphenols

Polyphenols are phytochemicals structurally composed of a phenol unit conjugated with organic acids or a sugar moiety. These molecules possess potency of several pharmacological activities including antioxidant, cell cytotoxicity through cell cycle arrest or apoptosis, estrogenic action, etc. These actions are related to their inherent

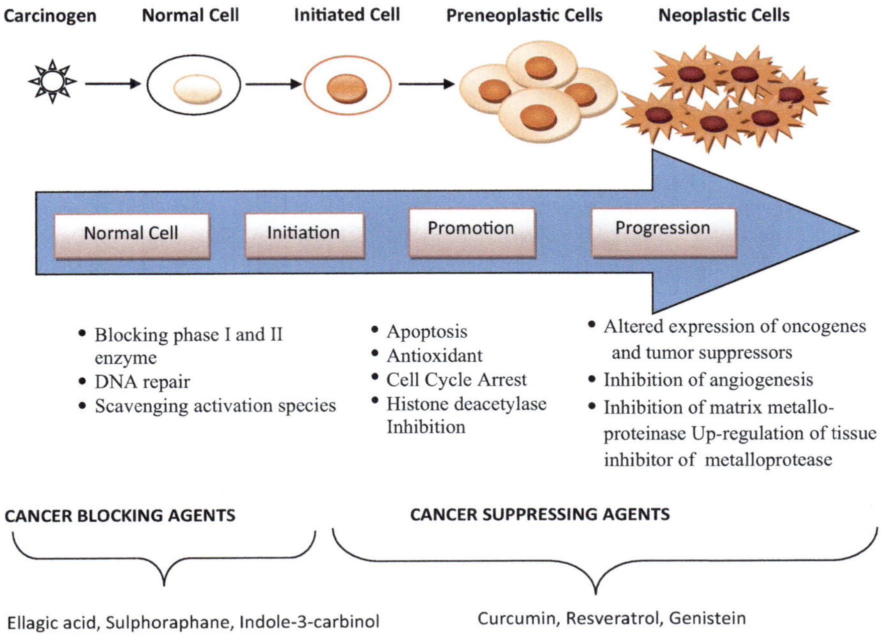

Fig. 2.2 Progression of cancer at different stages and involvement of phytochemicals to control molecular events

Table 2.3 Inflammation influencing signaling factors/pathways modulated by phytochemicals (Adopted from Zubair et al. 2017)

Signaling factor/ pathway	Phytochemical	References
COX2	Curcumin	Das and Vinayak (2015)
IL-1β	Resveratrol	Limagne et al. (2016)
IL-6	Resveratrol	Latruffe et al. (2015)
IL-8	Resveratrol	Latruffe et al. (2015)
iNOS	Curcumin	Das and Vinayak (2015)
TLR/IL-1R	Curcumin	Rana et al. (2016)
Keap1/Nrf2	Curcumin, EGCG, honokiol, plumbagin, resveratrol	Das and Vinayak (2015), Shanmugam et al. (2016), Kweon et al. (2006), Gao et al. (2016), Pan et al. (2015a, b), and Singh et al. (2014)
NF-kB	Curcumin, EGCG, honokiol, plumbagin, resveratrol	Kim et al. (2012), Pan et al. (2015a, b), Syed et al. (2007), Arora et al. (2012), Singh and Katiyar (2013), Ahmad et al. (2008), and Adhami et al. (2003)

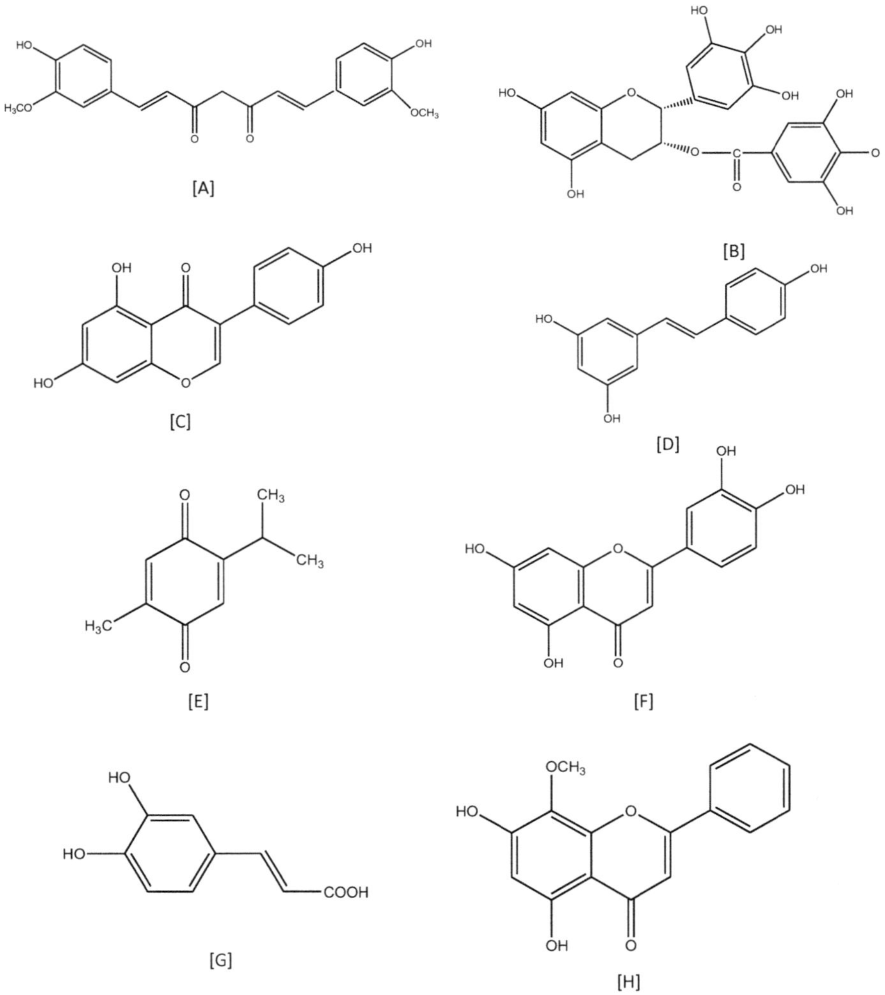

Fig. 2.3 Chemical structure of phytochemicals (**a**) Curcumin; (**b**) Epigallocatechin-3-gallate; (**c**) Genistein; (**d**) Resveratrol; (**e**) Thymoquinone; (**f**) Luteolin; (**g**) Caffeic acid; (**h**) Wogonin

chemopreventive potential. Polyphenols are known to act on multiple sites in the cellular system with several mechanism of action such as regulation of cell signaling, growth factors, cell survival, and apoptosis (Fresco et al. 2006). These phytochemicals are also involved in boosting the immune system so that it can distinguish and kill cancerous cell. Further, they inhibit angiogenesis, retard adhesion and invasion of cancer cell thereby reducing the chances of metastasis. Non-selective antioxidant action of polyphenols is additionally beneficial for chemoprevention (Tyagi et al. 2010). Inhibition of activation protein-1 (AP-1) needed for cell transformation, progression, and metastasis is related to polyphenol's chemopreventive activity. Further, activation of various signaling pathways including mitogen-activated protein kinase (MAPK) signaling pathways, c-Jun-terminal kinases (JNKs), stress activated protein kinases is also associated with their mechanism of action (Bode and Dong 2006).

2.2.1.1 Curcumin

Curcumin (CUR), a diferuloylmethane, is obtained from the spice turmeric and chemically it is (1,7-bis[4-hydroxy-3-methoxyphenyl]-1,6-heptadiene-3,5-dione) which belongs to polyphenol group (Aggarwal et al. 2007). CUR is a well-known compound enriched with chemopreventive, anti-inflammatory, anti-oxidative, and anti-carcinogenic properties (Kuo et al. 1996; Sa and Das 2008) along with anti-invasive and anti-metastatic potential. It is effective on number of cancer including breast, ovary, prostate, pancreas, and skin (Anand et al. 2008). It has been classified as third generation chemotherapeutic agent by National Cancer Institute. CUR acts by inhibiting proliferation and by inducing p53-dependent apoptosis of cancer cells (Li et al. 2005). It targets signaling pathways such as Notch 1 (Liao et al. 2011), chk1 (Sahu et al. 2009), COX-2 (Padhye et al. 2009), NF-κB, SP1 (Jutooru et al. 2010), and Stat3 (Glienke et al. 2009). Curcumin inhibits NF-κB pathway (Goel and Aggarwal 2010), c-Jun/AP-1 activation (Han et al. 2002), phosphorylation reactions (Liu et al. 1993), expression of matrix metalloproteases and COX-2 (Mohan et al. 2000). It acts as anti-invasive agent via epidermal growth factor receptor (EGFR) action (Korutla and Kumar 1994) inhibiting EGF kinase activity through the downregulation of NF-κB/AP-1 dependent metalloproteinase protein-1 (MMP) and -2 expression in breast cancer (Bachmeier et al. 2007). Despite its significant therapeutic potential, clinical applications of CUR are hindered due to very low systemic availability, degradation and metabolization, rapid systemic elimination leading to low bioavailability and pharmacological activity (Burgos-Morón et al. 2010). Thus, a very high dose up to 8 g is required for chemopreventive action (Sharma and Sukumar 2013).

2.2.2 Isothiocyanates (ITC)

The allyl ITC, benzyl ITC, phenethyl ITC, and sulforaphane (SFN) derived from garlic, broccoli, Brussels sprouts, cauliflower, cabbage, and similar cruciferous vegetables are used as chemopreventive agents belonging to the ITC group. These agents prevent carcinogen induced cancers. SFN, a naturally occurring sulfur-containing isothiocyanate, exhibits suppression of cancerous cell proliferation, G2-M phase cell cycle arrest and induces apoptosis by acting on several molecular targets such as signal transducer and Stat3, Akt, mitogen-activated protein kinase, p53, COX-2, NF-E2-related factor-2 (Kaminski et al. 2012). Such compounds reduce activity of NF-κB and have an effect on expression of adhesion molecules, inflammatory cytokines, and growth factors (Kallifatidis et al. 2009). SFN possesses large volume of distribution and remains in its active form at target site (Clarke et al. 2008).

2.2.2.1 Sulforaphane

Sulforaphane (SFN), which is abundantly available in cruciferous vegetables such as cauliflower and broccoli, etc., has potential to protect cells from oxidative damage and inflammation and hence is useful in chemoprevention of cancer. Recently, combined chemopreventive effect of aspirin, CUR, and free SFN after loading into a lipid based carrier system, i.e. solid lipid nanoparticles (SLN) for pancreatic cancer treatment has been reported (Sutaria et al. 2012). Author observed significantly reduced cell viability by 43.6 and 48.49% in MIAPaca-2 and Panc-1 cell lines when these drugs were given in combination of low doses. Alternatively an increased apoptosis of 61.3 and 60.37% can be achieved in MIA Paca-2 and Panc-1 cells after individual administration.

2.2.3 Flavonoids

Flavonoids, a group of polyphenolic compounds synthesized in plant cells, are promising chemopreventive agents for cancer. They can be obtained from fruits, vegetables, tea, wine, and cocoa (Harnly et al. 2006). Chemically, flavonoid consists of flavan nucleus where skeleton contains 15 carbon atom arranged in three rings (C6-C3-C6) and two aromatic rings are connected by a three carbon atom heterocyclic ring and an oxygen-containing pyran ring (Lu et al. 2013). Flavonoids may act as antioxidants (Pietta 2000) as free radical scavenging agents (Dugas Jr et al. 2000), anticancer agent, anti-inflammatory agents (Nijveldt et al. 2001), and enhance memory and learning power (Beking and Vieira 2010). Chemopreventive mechanisms of flavonoids include prevention of DNA damage and tumor promotion by free radical scavenging properties (Heijnen et al. 2001), regulation of oxidative stress and signaling pathways involved in carcinogenesis (Lee et al. 2010a, b). They also interfere with p53 mediated cell cycle progression (Plaumann et al. 1996), initiate apoptosis by activating caspase-9 and caspase-3 (Ren et al. 2003), modulate signaling pathway, and thus provide several molecular site for chemoprevention where conventional chemotherapeutic agents are ineffective (Amado et al. 2011). Flavanoids can be classified into several groups like flavonols (kaempferol, quercetin, and myricetin aglycones), flavan-3-ols (catechin epimers), flavanones (hesperetin and naringenin), flavones (apigenin and luteolin), isoflavones (phytoestrogens including soy genistein and daidzein), and anthocyanidins (pigments in the plant tegumental tissues) (Iriti 2011).

2.2.3.1 Anthocyanidins

Anthocyanidins, obtained from fruits and berries, have been shown to impediment cancer development in rodent models (Koide et al. 1996). Anthocyanins, anthocyanidins, their aglycons, especially cyanidin and delphinidin, have been widely studied for their chemopreventive mechanisms. In vitro experiments suggested their anti-oncogenic mechanisms that include anti-proliferation, induction of apoptosis, and inhibition of activities of oncogenic transcription factors and protein tyrosine kinases. The isomers of anthocyanins and anthocyanidins interconvert to each other depending on the pH, temperature, and presence of light. Anthocyanidins only survive for minutes in bioenvironment as they are very much prone to chemical decomposition and this must be considered in the sense of pharmaceutical drug development. Anthocyanidins present in berry, namely cyanidin, malvidin, peonidin, and delphinidin are widely explored for their chemopreventive and anticancer effects (Joshi and Goyal 2011). The combination therapy of these flavonoids works in synergistic manner to inhibit growth of cancer cells by effects on the oncogenic Notch and WNT pathways and their downstream targets (β-catenin, c-myc, cyclin D1, cyclin B1, pERK, MMP9, and VEGF proteins), enhanced cleavage of the apoptotic mediators Bcl2 and PARP and enhanced inhibition of TNFα-induced NF-kappa B activation.

2.2.3.2 Flavonols

Quercetin is most widely used flavonol and it is effective against number of cancers (Amado et al. 2009). It remains present in onion, apple, broccoli, and berries. Quercetin can be used as anticancer and chemopreventive agent as it inhibits cell growth by interfering cell cycle components (Lugli et al. 2009) and induces apoptosis and necrosis (Chen et al. 2010). Quercetin is now known to inhibit cell proliferation via G1 phase arrest and mitochondria mediated apoptosis, along with decreasing cell migration and invasion (Chen et al. 2013).

2.2.3.3 Flavones

Luteolin belongs to flavones group and one of the most effective flavonoids that demonstrate anticancer and chemopreventive effects. The chemopreventive effects of luteolin are mediated by affecting the signal transduction pathways, modulation of ROS levels, DNA replication enzymes (Manju and Nalini 2005), inhibition of topoisomerases I and II, reduction of NF-kB and AP-1 activity, stabilization of p53, and inhibition of PI3K, STAT3, IGF1R, and HER2 (López-Lázaro 2009). It also reduces the elevated level of lipid peroxides and cytochrome P450 (Elangovan et al. 1994).

2.2.3.4 Flavanones

Naringenin, a flavanone mostly found in citrus fruits, tomato, and grapefruit, is a representative example of this class of chemopreventive agents. It is known to have antioxidant activity, free radical scavenging nature, anti-inflammatory activity, and immune system modulator potential (Erlund 2004). Antiproliferative action of naringenin has been observed in many cancer cell lines including colon, breast, and uterus cancer cell lines. Its mechanism of action is closely related to antioxidant action, kinase and glucose uptake inhibition, ability to hamper cell proliferation via estrogen receptor (ER).

2.2.3.5 Catechins (Flavan-3-Ols)

Tea obtained from leaves of *Camellia sinensis* is a popular beverage, consumed in three forms green, black, and oolong tea (Graham 1992). The chemical constituents present in green tea are polyphenols and known as catechins include

(−)-epicatechin, (−)-epicatechin-3-gallate, (−)-epigallocatechin, and (−)-epigallocatechin-3-gallate (EGCG) (McKay and Blumberg 2002). Black tea contains the aflavins which are formed by enzymatic oxidation of polyphenols include TF-3-gallate, TF-3′-gallate, and TF-3-3′-digallate (Leone et al. 2003). These polyphenols inhibit cancers of different type. EGCG obtained from green tea has chemopreventive potential (Yang et al. 2002; Tammela et al. 2004) and it inhibits formation and development of tumors in different organs. EGCG acts on cancer cells in multi steps starting inhibition of cell proliferation via inducing cell apoptosis and cell cycle arrest, followed by modulation of transcription factor such as nuclear factor-κB (NF-κB) and activator protein (AP)-1 and finally through inhibition of cell invasion, angiogenesis, and metastasis by decreasing the production of matrix metalloproteinases (Yang et al. 2009). It also acts on reduction of oxidative stress and phase I enzymes along with induction of phase II enzyme activities (Kim et al. 2010).

2.2.3.6 Isoflavones

Genistein is an isoflavone present mainly in legumes. Genistein decreased the cell growth of cancer cells by inhibiting phosphorylation of extracellular signal-regulated kinase and Akt, involved in oral cancer proliferation (Johnson et al. 2010).

2.2.4 Phytoalexin

2.2.4.1 Resveratrol

Resveratrol present in grapes, berries, peanuts is a natural phytoalexin which possesses antioxidant (Pervaiz and Holme 2009), antiaging (Delmas et al. 2005), antiviral (Campagna and Rivas 2010), cardiovascular (Das and Das 2007), neuroprotective (Richard et al. 2011), and antiproliferative (Jang et al. 1997) properties. Resveratrol proved to be beneficial as anticancer, antiproliferative, and chemopreventive agent against various tumors including lymphoid (Dörrie et al. 2001), breast (Estrov et al. 2003), ovarian (Opipari et al. 2004), melanoma (Niles et al. 2003), and colon (Sengottuvelan et al. 2006). Anticancer action of resveratrol is mediated by interference of molecular signal transduction pathways that induce cancer cell death or inhibit cancer cell proliferation (She et al. 2001; Lee et al. 2011).

Despite such a broad spectrum of activity, the potential of resveratrol has restricted use due to poor solubility, instability, and low bioavailability (Signorelli and Ghidoni 2005). These limitations can be overcome by exploiting application of nanotechnology (Khushnud and Mousa 2013) that protect RSV from degradation, enhance bioavailability, and improve intracellular delivery (Sanna et al. 2012, 2013). Resveratrol modifies molecular pathways of tumor progression, such as Fas pathway, NF-κB and AP-1 transcriptional factors, and MAPK (Aggarwal et al. 2004) thus act as chemopreventive agent. It inhibits COX-1 (Li et al. 2012) and serves as a scavenger of radicals like hydroxyl and superoxide, etc. (Leonard et al. 2003). It has been regarded as antiproliferative agent due to pro-apoptotic potential caused by downregulation of a number of pro-proliferation and anti-apoptotic gene products (Ulrich et al. 2006).

2.2.5 Kaempferol

Kaempferol is more commonly found in apple, grapes, tomato, green tea, etc., as a secondary metabolite and belongs to flavonoid group. It is a chemopreventive agent that reduces the risk of ovarian cancer (Duthie and Crozier 2000), inhibits the expression of estrogen receptor in breast cancer cells (Hung 2004), and induces apoptosis by activation of MEK-MAPK (Sharma et al. 2007; Leung et al. 2007). Kaempferol also inhibits angiogenesis in ovarian cancer cells (Luo et al. 2012) and shows anti-inflammatory activity via inhibition of interleukin-4 (Cortes et al. 2007) and cyclooxygenase-2 expression by suppressing Src kinase (Lee et al. 2010a, b) and down-regulation of the NFκB cascade (García-Mediavilla et al. 2007).

2.3 Conclusion

Natural nutrients particularly phytochemicals serve as chemopreventive agents via diverse mechanisms and pathways. Many compounds like curcumin, resveratrol, EGCG, plumbagin, honokiol, genistein, naringenin, etc. which belong to different chemical classes are useful as chemopreventive agents against variety of cancers. Most of these compounds exert their action via multiple mechanism including induction of oxidative stress and suppression of NF-κB activation, p53-dependent and Fas-mediated pathways, induction of apoptosis and suppressing cell survival pathways, etc. The reported mechanisms are complex in nature; however clearly supports the chemopreventive activities shown by these natural phytoconstituents.

References

Adhami VM, Afaq F, Ahmad N (2003) Suppression of ultraviolet B exposure-mediated activation of NF-κB in normal human keratinocytes by resveratrol. Neoplasia 5(1):74

Agarwal B, Rao CV, Bhendwal S, Ramey WR, Shirin H, Reddy BS, Holt PR (1999) Lovastatin augments sulindac-induced apoptosis in colon cancer cells and potentiates chemopreventive effects of sulindac. Gastroenterology 117(4):838–847

Aggarwal BB, Bhardwaj A, Aggarwal RS, Seeram NP, Shishodia S, Takada Y (2004) Role of resveratrol in prevention and therapy of cancer: preclinical and clinical studies. Anticancer Res 24(5A):2783–2840

Aggarwal BB, Sundaram C, Malani N, Ichikawa H (2007) Curcumin: the Indian solid gold. In: The molecular targets and therapeutic uses of curcumin in health and disease. Springer, Boston, pp 1–75

Ahmad A, Banerjee S, Wang Z, Kong D, Sarkar FH (2008) Plumbagin-induced apoptosis of human breast cancer cells is mediated by inactivation of NF-κB and Bcl-2. J Cell Biochem 105 (6):1461–1471

Akshatha GM, Raval SK, Arpitha GM, Raval SH, Ghodasara DJ (2018) Immunohistochemical, histopathological study and chemoprotective effect of Solanum nigrum in N-nitrosodiethylamine-induced hepatocellular carcinoma in Wistar rats. Vet World 11(4):402

Amado NG, Cerqueira DM, Menezes FS, da Silva JFM, Neto VM, Abreu JG (2009) Isoquercitrin isolated from Hyptis fasciculata reduces glioblastoma cell proliferation and changes β-catenin cellular localization. Anti-Cancer Drugs 20(7):543–552

Amado NG, Fonseca BF, Cerqueira DM, Neto VM, Abreu JG (2011) Flavonoids: potential Wnt/beta-catenin signaling modulators in cancer. Life Sci 89(15-16):545–554

Anand P, Thomas SG, Kunnumakkara AB, Sundaram C, Harikumar KB, Sung B, Tharakan ST, Misra K, Priyadarsini IK, Rajasekharan KN, Aggarwal BB (2008) Biological activities of curcumin and its analogues (congeners) made by man and mother nature. Biochem Pharmacol 76(11):1590–1611

Arora S, Singh S, Piazza GA, Contreras CM, Panyam J, Singh AP (2012) Honokiol: a novel natural agent for cancer prevention and therapy. Curr Mol Med 12(10):1244–1252

Bachmeier B, Nerlich A, Iancu C, Cilli M, Schleicher E, Vené R, Dell'Eva R, Jochum M, Albini A, Pfeffer U (2007) The chemopreventive polyphenol Curcumin prevents hematogenous breast cancer metastases in immunodeficient mice. Cell Physiol Biochem 19(1-4):137–152

Beking K, Vieira A (2010) Flavonoid intake and disability-adjusted life years due to Alzheimer's and related dementias: a population-based study involving twenty-three developed countries. Public Health Nutr 13(9):1403–1409

Bode AM, Dong Z (2006) Molecular and cellular targets. Mol Carcinog 45(6):422–430

Bresalier RS, Sandler RS, Quan H, Bolognese JA, Oxenius B, Horgan K, Lines C, Riddell R, Morton D, Lanas A, Konstam MA (2005) Cardiovascular events associated with rofecoxib in a colorectal adenoma chemoprevention trial. N Engl J Med 352(11):1092–1102

Brisdelli F, D'Andrea G, Bozzi A (2009) Resveratrol: a natural polyphenol with multiple chemopreventive properties. Curr Drug Metab 10(6):530–546

Burgos-Morón E, Calderón-Montaño JM, Salvador J, Robles A, López-Lázaro M (2010) The dark side of curcumin. Int J Cancer 126(7):1771–1775

Campagna M, Rivas C (2010) Antiviral activity of resveratrol. Biochem Soc Trans 38(1):50–53

Chen C, Zhou J, Ji C (2010) Quercetin: a potential drug to reverse multidrug resistance. Life Sci 87 (11-12):333–338

Chen SF, Nien S, Wu CH, Liu CL, Chang YC, Lin YS (2013) Reappraisal of the anticancer efficacy of quercetin in oral cancer cells. J Chin Med Assoc 76(3):146–152

Chirumbolo S, Bjørklund G, Lysiuk R, Vella A, Lenchyk L, Upyr T (2018) Targeting cancer with phytochemicals via their fine tuning of the cell survival signaling pathways. Int J Mol Sci 19 (11):3568

Clark LC, Combs GF, Turnbull BW, Slate EH, Chalker DK, Chow J, Davis LS, Glover RA, Graham GF, Gross EG, Krongrad A (1996) Effects of selenium supplementation for cancer prevention in patients with carcinoma of the skin: a randomized controlled trial. JAMA 276 (24):1957–1963

Clarke JD, Dashwood RH, Ho E (2008) Multi-targeted prevention of cancer by sulforaphane. Cancer Lett 269(2):291–304

Cortes JR, Perez-G M, Rivas MD, Zamorano J (2007) Kaempferol inhibits IL-4-induced STAT6 activation by specifically targeting JAK3. J Immunol 179(6):3881–3887

Das S, Das DK (2007) Resveratrol: a therapeutic promise for cardiovascular diseases. Recent Pat Cardiovasc Drug Discov 2(2):133–138

Das L, Vinayak M (2015) Long term effect of curcumin in restoration of tumour suppressor p53 and phase-II antioxidant enzymes via activation of Nrf2 signalling and modulation of inflammation in prevention of cancer. PLoS One 10(4):e0124000

Delmas D, Jannin B, Latruffe N (2005) Resveratrol: preventing properties against vascular alterations and ageing. Mol Nutr Food Res 49(5):377–395

Deng C, Lu Q, Zhang Z, Rao T, Attwood J, Yung R, Richardson B (2003) Hydralazine may induce autoimmunity by inhibiting extracellular signal–regulated kinase pathway signaling. Arthritis Rheum 48(3):746–756

Desai AG, Qazi GN, Ganju RK, El-Tamer M, Singh J, Saxena AK, Bedi YS, Taneja SC, Bhat HK (2008) Medicinal plants and cancer chemoprevention. Curr Drug Metab 9(7):581–591

Dörrie J, Gerauer H, Wachter Y, Zunino SJ (2001) Resveratrol induces extensive apoptosis by depolarizing mitochondrial membranes and activating caspase-9 in acute lymphoblastic leukemia cells. Cancer Res 61(12):4731–4739

Dugas AJ Jr, Castañeda-Acosta J, Bonin GC, Price KL, Fischer NH, Winston GW (2000) Evaluation of the total peroxyl radical-scavenging capacity of flavonoids: structure—activity relationships. J Nat Prod 63(3):327–331

Duthie G, Crozier A (2000) Plant-derived phenolic antioxidants. Curr Opin Lipidol 11(1):43–47

Elangovan V, Sekar N, Govindasamy S (1994) Chemopreventive potential of dietary bioflavonoids against 20-methylcholanthrene-induced tumorigenesis. Cancer Lett 87(1):107–113

Erlund I (2004) Review of the flavonoids quercetin, hesperetin, and naringenin. Dietary sources, bioactivities, bioavailability, and epidemiology. Nutr Res 24(10):851–874

Estrov Z, Shishodia S, Faderl S, Harris D, Van Q, Kantarjian HM, Talpaz M, Aggarwal BB (2003) Resveratrol blocks interleukin-1beta-induced activation of the nuclear transcription factor NF-nB, inhibits proliferation, causes S-phase arrest, and induces apoptosis of acute myeloid leukemia cells. Blood 102(3):987–995

Fresco P, Borges F, Diniz C, Marques MPM (2006) New insights on the anticancer properties of dietary polyphenols. Med Res Rev 26(6):747–766

Gao DQ, Qian S, Ju T (2016) Anticancer activity of honokiol against lymphoid malignant cells via activation of ROS-JNK and attenuation of Nrf2 and NF-κB. J BUON 21(3):673–679

García-Mediavilla V, Crespo I, Collado PS, Esteller A, Sánchez-Campos S, Tuñón MJ, González-Gallego J (2007) The anti-inflammatory flavones quercetin and kaempferol cause inhibition of inducible nitric oxide synthase, cyclooxygenase-2 and reactive C-protein, and down-regulation of the nuclear factor kappaB pathway in Chang liver cells. Eur J Pharmacol 557(2-3):221–229

Glienke W, Maute L, Wicht J, Bergmann L (2009) Curcumin inhibits constitutive STAT3 phosphorylation in human pancreatic cancer cell lines and downregulation of survivin/BIRC5 gene expression. Cancer Investig 28(2):166–171

Goel A, Aggarwal BB (2010) Curcumin, the golden spice from Indian saffron, is a chemosensitizer and radiosensitizer for tumors and chemoprotector and radioprotector for normal organs. Nutr Cancer 62(7):919–930

Graham HN (1992) Green tea composition, consumption, and polyphenol chemistry. Prev Med 21 (3):334–350

Grandhi BK, Thakkar A, Wang J, Prabhu S (2013) A novel combinatorial nanotechnology-based oral chemopreventive regimen demonstrates significant suppression of pancreatic cancer neoplastic lesions. Cancer Prev Res 6(10):1015–1025

Han SS, Keum YS, Seo HJ, Surh YJ (2002) Curcumin suppresses activation of NF-kappaB and AP-1 induced by phorbol ester in cultured human promyelocytic leukemia cells. J Biochem Mol Biol 35(3):337–342

Harnly JM, Doherty RF, Beecher GR, Holden JM, Haytowitz DB, Bhagwat S, Gebhardt S (2006) Flavonoid content of US fruits, vegetables, and nuts. J Agric Food Chem 54(26):9966–9977

Heijnen CGM, Haenen GRMM, Van Acker FAA, Van der Vijgh WJF, Bast A (2001) Flavonoids as peroxynitrite scavengers: the role of the hydroxyl groups. Toxicol In Vitro 15(1):3–6

Hung H (2004) Inhibition of estrogen receptor alpha expression and function in MCF-7 cells by kaempferol. J Cell Physiol 198(2):197–208

Iriti M (2011) Editorial [hot topic: introduction to polyphenols, plant chemicals for human health (guest editor: Marcello Iriti)]. Mini Rev Med Chem 11(14):1183–1185

Iriti M, Faoro F (2009a) Bioactivity of grape chemicals for human health. Nat Prod Commun 4 (5):611–634

Iriti M, Faoro F (2009b) Chemical diversity and defence metabolism: how plants cope with pathogens and ozone pollution. Int J Mol Sci 10(8):3371–3399

Jang M, Cai L, Udeani GO, Slowing KV, Thomas CF, Beecher CW, Fong HH, Farnsworth NR, Kinghorn AD, Mehta RG, Moon RC (1997) Cancer chemopreventive activity of resveratrol, a natural product derived from grapes. Science 275(5297):218–220

Johnson TL, Lai MB, Lai JC, Bhushan A (2010) Inhibition of cell proliferation and MAP kinase and Akt pathways in oral squamous cell carcinoma by genistein and biochanin A. Evid Based Complement Alternat Med 7

Joshi Y, Goyal B (2011) Anthocyanins: a lead for anticancer drugs. Int J Res Pharm 1 (4):1119–1126

Jutooru I, Chadalapaka G, Lei P, Safe S (2010) Inhibition of NFκB and pancreatic cancer cell and tumor growth by curcumin is dependent on specificity protein down-regulation. J Biol Chem 285(33):25332–25344

Kallifatidis G, Rausch V, Baumann B, Apel A, Beckermann BM, Groth A, Mattern J, Li Z, Kolb A, Moldenhauer G, Altevogt P (2009) Sulforaphane targets pancreatic tumour-initiating cells by NF-κB-induced antiapoptotic signalling. Gut 58(7):949–963

Khushnud T, Mousa SA (2013) Potential role of naturally derived polyphenols and their nanotechnology delivery in cancer. Mol Biotechnol 55(1):78–86

Kim JW, Amin AR, Shin DM (2010) Chemoprevention of head and neck cancer with green tea polyphenols. Cancer Prev Res 3(8):900–909

Kim JH, Gupta SC, Park B, Yadav VR, Aggarwal BB (2012) Turmeric (Curcuma longa) inhibits inflammatory nuclear factor (NF)-κB and NF-κB-regulated gene products and induces death receptors leading to suppressed proliferation, induced chemosensitization, and suppressed osteoclastogenesis. Mol Nutr Food Res 56(3):454–465

Koide T, Kamei H, Hashimoto Y, Kojima T, Hasegawa M (1996) Antitumor effect of hydrolyzed anthocyanin from grape rinds and red rice. Cancer Biother Radiopharm 11(4):273–277

Komakech R, Kang Y, Lee JH, Omujal F (2017) A review of the potential of phytochemicals from Prunus africana (Hook f.) Kalkman stem bark for chemoprevention and chemotherapy of prostate cancer. Evid Based Complement Alternat Med 2017

Korutla L, Kumar R (1994) Inhibitory effect of curcumin on epidermal growth factor receptor kinase activity in A431 cells. Biochim Biophys Acta 1224(3):597–600

Kuo PL, Lin CC (2003) Green tea constituent (−)-epigallocatechin-3-gallate inhibits Hep G2 cell proliferation and induces apoptosis through p53-dependent and Fas-mediated pathways. J Biomed Sci 10(2):219–227

Kuo ML, Huang TS, Lin JK (1996) Curcumin, an antioxidant and anti-tumor promoter, induces apoptosis in human leukemia cells. Biochim Biophys Acta 1317(2):95–100

Kweon MH, Adhami VM, Lee JS, Mukhtar H (2006) Constitutive overexpression of Nrf2-dependent heme oxygenase-1 in A549 cells contributes to resistance to apoptosis induced by epigallocatechin 3-gallate. J Biol Chem 281(44):33761–33772

Latruffe N, Lançon A, Frazzi R, Aires V, Delmas D, Michaille JJ, Djouadi F, Bastin J, Cherkaoui-Malki M (2015) Exploring new ways of regulation by resveratrol involving miRNAs, with emphasis on inflammation. Ann N Y Acad Sci *1348*(1):97–106

Lee KM, Lee KW, Jung SK, Lee EJ, Heo YS, Bode AM, Lubet RA, Lee HJ, Dong Z (2010a) Kaempferol inhibits UVB-induced COX-2 expression by suppressing Src kinase activity. Biochem Pharmacol 80(12):2042–2049

Lee DE, Shin BJ, Hur HJ, Kim JH, Kim J, Kang NJ, Kim DO, Lee CY, Lee KW, Lee HJ (2010b) Quercetin, the active phenolic component in kiwifruit, prevents hydrogen peroxide-induced inhibition of gap-junction intercellular communication. Br J Nutr 104(2):164–170

Lee KW, Bode AM, Dong Z (2011) Molecular targets of phytochemicals for cancer prevention. Nat Rev Cancer 11(3):211–218

Leonard SS, Xia C, Jiang BH, Stinefelt B, Klandorf H, Harris GK, Shi X (2003) Resveratrol scavenges reactive oxygen species and effects radical-induced cellular responses. Biochem Biophys Res Commun 309(4):1017–1026

Leone M, Zhai D, Sareth S, Kitada S, Reed JC, Pellecchia M (2003) Cancer prevention by tea polyphenols is linked to their direct inhibition of antiapoptotic Bcl-2-family proteins. Cancer Res 63(23):8118–8121

Leung HWC, Lin CJ, Hour MJ, Yang WH, Wang MY, Lee HZ (2007) Kaempferol induces apoptosis in human lung non-small carcinoma cells accompanied by an induction of antioxidant enzymes. Food Chem Toxicol 45(10):2005–2013

Li DWC, Liu JP, Mao YW, Xiang H, Wang J, Ma WY, Dong Z, Pike HM, Brown RE, Reed JC (2005) Calcium-activated RAF/MEK/ERK signaling pathway mediates p53-dependent apoptosis and is abrogated by αB-crystallin through inhibition of RAS activation. Mol Biol Cell 16 (9):4437–4453

Li H, Xia N, Förstermann U (2012) Cardiovascular effects and molecular targets of resveratrol. Nitric Oxide 26(2):102–110

Liao S, Xia J, Chen Z, Zhang S, Ahmad A, Miele L, Sarkar FH, Wang Z (2011) Inhibitory effect of curcumin on oral carcinoma CAL-27 cells via suppression of Notch-1 and NF-κB signaling pathways. J Cell Biochem 112(4):1055–1065

Limagne E, Lançon A, Delmas D, Cherkaoui-Malki M, Latruffe N (2016) Resveratrol interferes with IL1-β-induced pro-inflammatory paracrine interaction between primary chondrocytes and macrophages. Nutrients 8(5):280

Liu JY, Lin SJ, Lin JK (1993) Inhibitory effects of curcumin on protein kinase C activity induced by 12-O-tetradecanoyl-phorbol-13-acetate in NIH 3T3 cells. Carcinogenesis 14(5):857–861

López-Lázaro M (2009) Distribution and biological activities of the flavonoid luteolin. Mini Rev Med Chem 9(1):31–59

Lu MF, Xiao ZT, Zhang HY (2013) Where do health benefits of flavonoids come from? Insights from flavonoid targets and their evolutionary history. Biochem Biophys Res Commun 434 (4):701–704

Lucenteforte E, Garavello W, Bosetti C, La Vecchia C (2009) Dietary factors and oral and pharyngeal cancer risk. Oral Oncol 45(6):461–467

Lugli E, Ferraresi R, Roat E, Troiano L, Pinti M, Nasi M, Nemes E, Bertoncelli L, Gibellini L, Salomoni P, Cooper EL (2009) Quercetin inhibits lymphocyte activation and proliferation without inducing apoptosis in peripheral mononuclear cells. Leuk Res 33(1):140–150

Luo H, Rankin GO, Juliano N, Jiang BH, Chen YC (2012) Kaempferol inhibits VEGF expression and in vitro angiogenesis through a novel ERK-NFκB-cMyc-p21 pathway. Food Chem 130 (2):321–328

Kaminski BM, Steinhilber D, Stein JM, Ulrich S (2012) Phytochemicals resveratrol and sulforaphane as potential agents for enhancing the anti-tumor activities of conventional cancer therapies. Curr Pharm Biotechnol 13(1):137–146

Majumdar D, Jung KH, Zhang H, Nannapaneni S, Wang X, Amin AR, Chen Z, Shin DM (2014) Luteolin nanoparticle in chemoprevention: in vitro and in vivo anticancer activity. Cancer Prev Res 7(1):65–73

Manju V, Nalini N (2005) Chemopreventive potential of luteolin during colon carcinogenesis induced by 1, 2-dimethylhydrazine. Ital J Biochem 54(3-4):268–275

Marks PA, Richon VM, Rifkind RA (2000) Histone deacetylase inhibitors: inducers of differentiation or apoptosis of transformed cells. J Natl Cancer Inst 92(15):1210–1216

McKay DL, Blumberg JB (2002) The role of tea in human health: an update. J Am Coll Nutr 21 (1):1–13

Mehta RG, Williamson E, Patel MK, Koeffler HP (2000) A ligand of peroxisome proliferator-activated receptor γ, retinoids, and prevention of preneoplastic mammary lesions. J Natl Cancer Inst 92(5):418–423

Mohan R, Sivak J, Ashton P, Russo LA, Pham BQ, Kasahara N, Raizman MB, Fini ME (2000) Curcuminoids inhibit the angiogenic response stimulated by fibroblast growth factor-2, including expression of matrix metalloproteinase gelatinase B. J Biol Chem 275(14):10405–10412

Nijveldt RJ, Van Nood ELS, Van Hoorn DE, Boelens PG, Van Norren K, Van Leeuwen PA (2001) Flavonoids: a review of probable mechanisms of action and potential applications. Am J Clin Nutr 74(4):418–425

Niles RM, McFarland M, Weimer MB, Redkar A, Fu YM, Meadows GG (2003) Resveratrol is a potent inducer of apoptosis in human melanoma cells. Cancer Lett 190(2):157–163

Nussmeier NA, Whelton AA, Brown MT, Langford RM, Hoeft A, Parlow JL, Boyce SW, Verburg KM (2005) Complications of the COX-2 inhibitors parecoxib and valdecoxib after cardiac surgery. N Engl J Med 352(11):1081–1091

Opipari AW, Tan L, Boitano AE, Sorenson DR, Aurora A, Liu JR (2004) Resveratrol-induced autophagocytosis in ovarian cancer cells. Cancer Res 64(2):696–703

Padhye S, Banerjee S, Chavan D, Pandye S, Swamy KV, Ali S, Li J, Dou QP, Sarkar FH (2009) Fluorocurcumins as cyclooxygenase-2 inhibitor: molecular docking, pharmacokinetics and tissue distribution in mice. Pharm Res 26(11):2438–2445

Pan H, Chen J, Shen K, Wang X, Wang P, Fu G, Meng H, Wang Y, Jin B (2015a) Mitochondrial modulation by Epigallocatechin 3-Gallate ameliorates cisplatin induced renal injury through decreasing oxidative/nitrative stress, inflammation and NF-kB in mice. PLoS One 10(4): e0124775

Pan ST, Qin Y, Zhou ZW, He ZX, Zhang X, Yang T, Yang YX, Wang D, Zhou SF, Qiu JX (2015b) Plumbagin suppresses epithelial to mesenchymal transition and stemness via inhibiting Nrf2-mediated signaling pathway in human tongue squamous cell carcinoma cells. Drug Des Devel Ther 9:5511

Pervaiz S, Holme AL (2009) Resveratrol: its biologic targets and functional activity. Antioxid Redox Signal 11(11):2851–2897

Pietta PG (2000) Flavonoids as antioxidants. J Nat Prod 63(7):1035–1042

Plaumann B, Fritsche M, Rimpler H, Brandner G, Hess RD (1996) Flavonoids activate wild-type p53. Oncogene 13(8):1605–1614

Prasad NR, Karthikeyan A, Karthikeyan S, Reddy BV (2011) Inhibitory effect of caffeic acid on cancer cell proliferation by oxidative mechanism in human HT-1080 fibrosarcoma cell line. Mol Cell Biochem 349(1-2):11–19

Rana M, Maurya P, Reddy SS, Singh V, Ahmad H, Dwivedi AK, Dikshit M, Barthwal MK (2016) A standardized chemically modified Curcuma longa extract modulates IRAK-MAPK signaling in inflammation and potentiates cytotoxicity. Front Pharmacol 7:223

Ren W, Qiao Z, Wang H, Zhu L, Zhang L (2003) Flavonoids: promising anticancer agents. Med Res Rev 23(4):519–534

Richard T, Pawlus AD, Iglésias ML, Pedrot E, Waffo-Teguo P, Mérillon JM, Monti JP (2011) Neuroprotective properties of resveratrol and derivatives. Ann N Y Acad Sci 1215(1):103

Rudin CM, Cohen EE, Papadimitrakopoulou VA, Silverman S Jr, Recant W, El-Naggar AK, Stenson K, Lippman SM, Hong WK, Vokes EE (2003) An attenuated adenovirus, ONYX-015, as mouthwash therapy for premalignant oral dysplasia. J Clin Oncol Off J Am Soc Clin Oncol 21(24):4546

Sa G, Das T (2008) Anti cancer effects of curcumin: cycle of life and death. Cell Div 3(1):14

Sahu RP, Batra S, Srivastava SK (2009) Activation of ATM/Chk1 by curcumin causes cell cycle arrest and apoptosis in human pancreatic cancer cells. Br J Cancer 100(9):1425–1433

Salim LZA, Mohan S, Othman R, Abdelwahab SI, Kamalidehghan B, Sheikh BY, Ibrahim MY (2013) Thymoquinone induces mitochondria-mediated apoptosis in acute lymphoblastic leukaemia in vitro. Molecules 18(9):11219–11240

Sanna V, Roggio AM, Siliani S, Piccinini M, Marceddu S, Mariani A, Sechi M (2012) Development of novel cationic chitosan-and anionic alginate–coated poly (d, l-lactide-co-glycolide) nanoparticles for controlled release and light protection of resveratrol. Int J Nanomedicine 7:5501

Sanna V, Siddiqui IA, Sechi M, Mukhtar H (2013) Resveratrol-loaded nanoparticles based on poly (epsilon-caprolactone) and poly (d, l-lactic-co-glycolic acid)–poly (ethylene glycol) blend for prostate Cancer treatment. Mol Pharm 10(10):3871–3881

Sengottuvelan M, Viswanathan P, Nalini N (2006) Chemopreventive effect of trans-resveratrol-a phytoalexin against colonic aberrant crypt foci and cell proliferation in 1, 2-dimethylhydrazine induced colon carcinogenesis. Carcinogenesis 27(5):1038–1046

Shanmugam T, Selvaraj M, Poomalai S (2016) Epigallocatechin gallate potentially abrogates fluoride induced lung oxidative stress, inflammation via Nrf2/Keap1 signaling pathway in rats: an in-vivo and in-silico study. Int Immunopharmacol 39:128–139

Sharma D, Sukumar S (2013) Big punches come in nanosizes for chemoprevention. Cancer Prev Res 6(10):1007–1010

Sharma V, Joseph C, Ghosh S, Agarwal A, Mishra MK, Sen E (2007) Kaempferol induces apoptosis in glioblastoma cells through oxidative stress. Mol Cancer Ther 6(9):2544–2553

She QB, Bode AM, Ma WY, Chen NY, Dong Z (2001) Resveratrol-induced activation of p53 and apoptosis is mediated by extracellular-signal-regulated protein kinases and p38 kinase. Cancer Res 61(4):1604–1610

Shin DM, Ro JY, Hong WK, Hittelman WN (1994) Dysregulation of epidermal growth factor receptor expression in premalignant lesions during head and neck tumorigenesis. Cancer Res 54 (12):3153–3159

Signorelli P, Ghidoni R (2005) Resveratrol as an anticancer nutrient: molecular basis, open questions and promises. J Nutr Biochem 16(8):449–466

Singh T, Katiyar SK (2013) Honokiol inhibits non-small cell lung cancer cell migration by targeting PGE2-mediated activation of β-catenin signaling. PLoS One 8(4):e60749

Singh B, Shoulson R, Chatterjee A, Ronghe A, Bhat NK, Dim DC, Bhat HK (2014) Resveratrol inhibits estrogen-induced breast carcinogenesis through induction of NRF2-mediated protective pathways. Carcinogenesis 35(8):1872–1880

Sutaria D, Grandhi BK, Thakkar A, Wang J, Prabhu S (2012) Chemoprevention of pancreatic cancer using solid-lipid nanoparticulate delivery of a novel aspirin, curcumin and sulforaphane drug combination regimen. Int J Oncol 41(6):2260–2268

Syed DN, Afaq F, Kweon MH, Hadi N, Bhatia N, Spiegelman VS, Mukhtar H (2007) Green tea polyphenol EGCG suppresses cigarette smoke condensate-induced NF-κ B activation in normal human bronchial epithelial cells. Oncogene 26(5):673–682

Tammela P, Ekokoski E, García-Horsman A, Talman V, Finel M, Tuominen R, Vuorela P (2004) Screening of natural compounds and their derivatives as potential protein kinase C inhibitors. Drug Dev Res 63(2):76–87

Tan W, Lu J, Huang M, Li Y, Chen M, Wu G, Gong J, Zhong Z, Xu Z, Dang Y, Guo J (2011) Anti-cancer natural products isolated from Chinese medicinal herbs. Chin Med 6(1):27

Thompson PA, Wertheim BC, Zell JA, Chen WP, McLaren CE, LaFleur BJ, Meyskens FL, Gerner EW (2010) Levels of rectal mucosal polyamines and prostaglandin E2 predict ability of DFMO and sulindac to prevent colorectal adenoma. Gastroenterology 139(3):797–805

Torrance CJ, Jackson PE, Montgomery E, Kinzler KW, Vogelstein B, Wissner A, Nunes M, Frost P, Discafani CM (2000) Combinatorial chemoprevention of intestinal neoplasia. Nat Med 6(9):1024–1028

Tyagi S, Singh G, Sharma A, Aggarwal G (2010) Phytochemicals as candidate therapeutics: an overview. Int J Pharm Sci Rev Res 3(1):53–55

Ulrich S, Loitsch SM, Rau O, von Knethen A, Brüne B, Schubert-Zsilavecz M, Stein JM (2006) Peroxisome proliferator–activated receptor γ as a molecular target of resveratrol-induced modulation of polyamine metabolism. Cancer Res 66(14):7348–7354

Xu Q, Zong L, Chen X, Jiang Z, Nan L, Li J, Duan W, Lei J, Zhang L, Ma J, Li X (2015) Resveratrol in the treatment of pancreatic cancer. Ann N Y Acad Sci 1348(1):10–19

Yang CS, Maliakal P, Meng X (2002) Inhibition of carcinogenesis by tea. Annu Rev Pharmacol Toxicol 42(1):25–54

Yang CS, Wang X, Lu G, Picinich SC (2009) Cancer prevention by tea: animal studies, molecular mechanisms and human relevance. Nat Rev Cancer 9(6):429–439

Zhang LL, Li L, Wu DP, Fan JH, Li X, Wu KJ, Wang XY, He DL (2008) A novel anti-cancer effect of genistein: reversal of epithelial mesenchymal transition in prostate cancer cells 1. Acta Pharmacol Sin 29(9):1060–1068

Zubair H, Azim S, Ahmad A, Khan MA, Patel GK, Singh S, Singh AP (2017) Cancer chemoprevention by phytochemicals: nature's healing touch. Molecules 22(3):395

Controlled Delivery of Chemopreventive Agents

3

Abstract

Phytoconstituents have some limitations like poor bioavailability, stability issues and hence limited therapeutic efficacy upon administration. Such issues can be warranted using the controlled delivery of these molecules with the help of novel drug delivery systems especially nano-sized carrier systems. A wide range of biomaterial and their suitable engineering into a nanocarrier navigate them to the target site. In this chapter, we have covered the aspects of various nanosystems that are being used for the delivery of chemopreventive agents like polymeric nanoparticles, solid lipid nanoparticles, vesicular systems (liposomes and phytosomes), metallic nanoparticles (iron, gold, silver, and silica), nanocapsules, and hybrid nanosystems.

Keywords

Controlled drug delivery · Phytochemicals · Anticancer drug · Novel drug delivery system · Chemoprevention

3.1 Introduction

Although a large of number of phytochemicals are known to have excellent pharmacological activities, however, they are lacking in many translational issues at pre-clinical and clinical level. Most of the compounds have low aqueous solubility and therefore poor bioavailability in plasma pool. Some of the phytochemicals have high metabolism, chemical degradation during en routing bioenvironment, and rapid renal clearance as well. All these micro-events play a critical role in ultimate inadequate concentration of phytochemicals at the target site; this leads to poor therapeutic action and sometimes no action as well (Bharali et al. 2011). In order to deal with issues like stability in biological environment, low bioavailability, dose

R. Paliwal, S. R. Paliwal, *Advances in Nanochemoprevention*,
https://doi.org/10.1007/978-981-15-9692-6_3

inaccuracy, and desired release pattern, the phytoconstituents have been either encapsulated or entrapped into various types of drug delivery systems (Ranjan et al. 2013). Additionally, such delivery system can be grafted with the ligands to achieve targeted therapy especially in case of treatment of variety of cancer.

3.2 Controlled Delivery of Phytoconstituents: Advantages and Limitations

Controlled phytoconstituents delivery has been achieved using both micro-particulate system as well as nano-sized system. Both hydrophobic and hydrophilic moieties can be loaded in the matrix of the polymer or lipid else can be trapped into the aqueous environment of vesicular systems like liposome, niosomes, and phytosomes, etc. (Paliwal et al. 2012a, b, c; Rai et al. 2008). The encapsulation of phytoconstituents in to a suitable carrier system provides opportunities to modulate the trafficking of the loaded active chemicals towards the target site to exert its therapeutic effect (Rajalakshmi et al. 2018). For example, Berberine is a drug of choice for lymphatic disorders and it should reach into lymphatic system. For that such drugs or phytoconstituents can be encapsulated into a lipid based drug carrier system so that it can follow chylomicron mimicking to achieve higher concentration into the desired location and avoid unnecessary first pass metabolism (Elsheikh et al. 2018). It is noted here that some of the commercial products are available such as Lipocurc® (liposomal formulation for intravenous route) and Meriva® which is a phytosomal mixture of curcumminoid with phospholipid lecithin (Bolger et al. 2019; Belcaro et al. 2010). Such formulations have successfully enhanced the bioavailability of curcumin in clinical settings. However, few limitations of controlled delivery are indispensible till date like overall time and effort in developing successful products, scale-up and regulatory issues of nanomedicine, commercial viability, and tech transfer of the lab work into clinical reality (Paliwal et al. 2014; Hua et al. 2018). The compatibility of the ingredients and the desired drug release correlating with in vitro behavior to in vivo performance are also some of the issues. Table 3.1 summarizes the advantages and disadvantages of both conventional and nanoformulations.

3.2.1 Nanoformulation with Respect to Cancer Cell Physiology and Biology

The desired characteristics of a nanoformulation for achieving higher therapeutic concentration at cancer cell target are particle size (preferably less than 100 nm), stealthing character to avoid opsonization by reticuloendothelial system (RES), uniform size distribution, high drug payload, target sensitivity, and target selectivity (Lagoa et al. 2020). Tumor vasculature is leaky in nature and therefore due to enhance permeability and retention effect, nano-sized particles/carriers get into the tumor vicinity and accumulated there as a result of passive targeting (Paliwal et al.

Table 3.1 Advantages and disadvantages of conventional and nanoformulations (Adopted from Davatgaran-Taghipour et al. 2017)

Nanoformulations		Conventional formulations	
Advantages	Disadvantages	Advantages	Disadvantages
• Higher surface area to volume ratio. • Improved bioavailability. • Targeted drug delivery. • Sustained drug release. • Protection from environment.	• Short shelf life. • Unpredictable toxicity, stability, and pharmacokinetics. • More expensive.	• Specified safety profile. • Predicted toxicity, stability and pharmacokinetics. • Less expensive.	• Untargeted drug delivery. • Uncontrolled release. • Undesirable pharmacokinetics.

2011a, b). Alternatively, nanoparticles surface could be modified using a suitable ligand for cancerous cell receptors that are over-expressed for active targeting though receptor-mediated internalization (Paliwal et al. 2012a, b, c). Drug–polymer or drug–lipid conjugates can also be developed to achieve the similar outcomes in cancer therapy. Utilization of tumor microenvironment is also one of the approaches for controlled release of the cytotoxic/chemopreventive drug. Here, the nanoparticles properties to release the drug in low pH environment or in presence of enzymes are explored. Intracellular release of the chemopreventive agents could also be achieved using cationic nanoparticles or ligand modified nanosystems (Paliwal et al. 2016).

3.3 Role of Physiochemical Properties of Phytoconstituents/ Chemopreventive Drugs in Formulation Design

As described above, most of the phytoconstituents or chemopreventive agents are hydrophobic in nature and therefore they require suitable engineered system for better drug loading and desired drug release (Khan and Gurav 2018). Furthermore, solubility, partition coefficient, pK_a, molecular weight, polymorphism are also important critical features to decide formulation design and strategy. For example, lipid based carrier system is chosen for hydrophobic drugs; however, process of development of carrier system should be compatible with the nature of the drug and excipients selected for the development as they ultimately affect the overall performance of the integrated system (Katdare and Chaubal 2006).

3.4 Biomaterial Used for Controlled Delivery of Bioactives

Apart from physicochemical characteristics of the chemopreventive agent, nature of biomaterial selected for nanoparticles synthesis or formulation is also crucial as it determines the fate of the controlled delivery at large. A range of different class and sub-class of biomaterial including polymers (natural or synthetic or semi-synthetic), lipids (solid lipid or phospholipids), proteins (animal proteins or plants proteins), polysaccharides, metals, etc., have been tested and reported for controlled release of chemopreventive phytoconstituents.

3.5 Nanocarriers Used for Controlled Delivery of Phytoactives

Different types of nanocarriers, i.e. particulate systems (nanoparticles or nanocapsules), vesicular systems (liposomes, niosomes, or phytosomes), self-assembling system (micelles), lipid based emulsions, metallic nanoparticles (gold nanoparticles or silver nanoparticles), and multifunctional systems (ligand anchored stealthing system) have been reported in literature for controlled delivery of phytoactives (Martínez-Ballesta et al. 2018; Subramanian et al. 2016). Figure 3.1

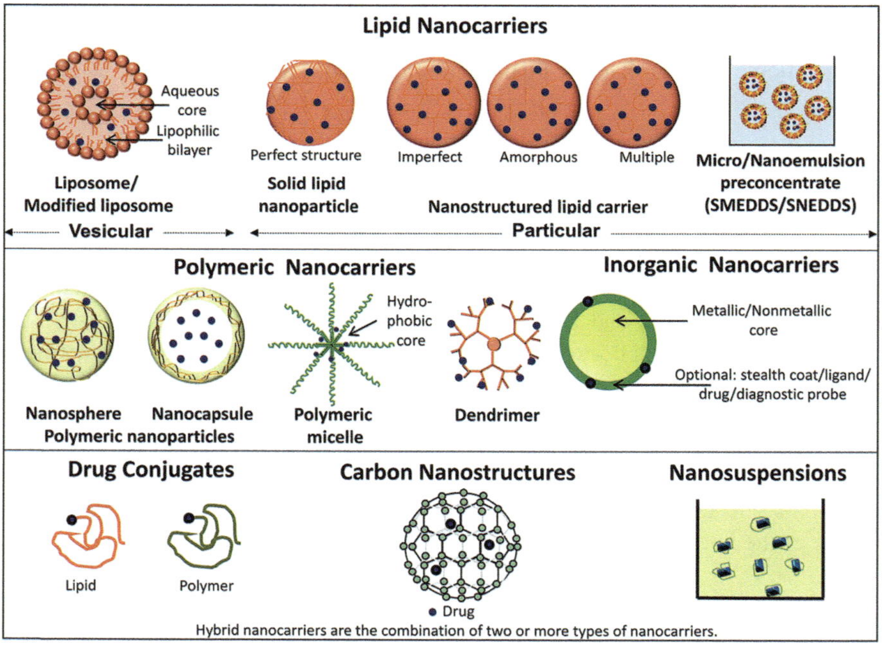

Fig. 3.1 Schematic representation depicting types of nanocarriers (Adopted from the reference Desai et al. 2020)

represents various types of nanocarriers for their possible use in nanochemoprevention.

3.5.1 Polymeric Nanoparticles

Nanoparticles composed of both natural and synthetic polymers have been utilized for diverse phytoconstituents. PLGA, PCL, Eudragit, PEG based nanosystems have been formulated and designed for controlled drug delivery (Elsabahy and Wooley 2012). A safe and green status of PLGA makes it an appropriate biomaterial for most of the compounds and therefore is preferred choice of formulation scientist. Formulation development approaches of polymeric nanoparticles can be classified into two parts; (a) top-down approaches and (b) bottom-up approaches (Ezhilarasi et al. 2013). Top-down approaches include energy inputs and are acceptable by the industry; however, these approaches are not suitable for thermosensitive drugs and are costly for small laboratory batches. Bottom-up approaches involve use of organic solvents and have regulatory issues at the time of translation from bench to market.

3.5.2 Solid Lipid Nanoparticles

Solid lipid based nanoparticles are composed of inner core which is made up of solid lipids (mono-, di-, or tri-glycerides or their combinations) and a surfactant or stabilizer (phospholipids, non-ionic surfactant, etc.) (Paliwal et al. 2009). SLNs as carriers provide opportunity to develop a system comprised of more compatible biomaterials than polymers in term of acceptance by regulatory agencies (Paliwal et al. 2011a, b). Some of the methods of SLNs development are industry ready which can turn up a laboratory success into real industrial outcomes. For more distorted type of matrix, nanostructured lipid carriers (NLCs) can be adopted as they are devoid of drug expulsion problems like SLNs.

3.5.3 Vesicular Systems

Vesicle based carrier systems offer opportunity to incorporate both lipophilic phytodrugs (among bilayers) and hydrophilic phytodrugs (inside the aqueous compartment). Depending upon the composition of the excipients selected for formulation development, these systems can be further classified into liposomes (phospholipid based vesicles), niosomes (non-ionic surfactant based system), and phytosomes (phytophospholipid based system), etc. (Jain et al. 2014).

3.5.3.1 Liposomes

Liposomes are phospholipids based vesicular system composed of inner aqueous compartment enclosed with phospholipids bilayers. Commercial successful liposomal products for controlled release of bioactive are available, namely

DOXIL, MYOCET, AMBISOME, etc. Liposomes can be easily modified for desired characteristics like magnetic liposomes, pH-sensitive liposomes, targeted liposomes, stealth liposomes, and surface modified liposomes for variety of purposes of drug delivery domain (Paliwal et al. 2012a, b, c). They can be delivered through almost all routes of administration, i.e. oral, topical, parenteral, ocular, etc. As described earlier, curcumin loaded liposomal product LIPOCURC® is commercially available as well. This demonstrates the benefits of the liposomes as drug carrier for phytoconstituents (Singh et al. 2019).

3.5.3.2 Phytosomes

Phytosomes represent another class of vesicular systems that is composed of phospholipids and is quite similar to liposomes in compositions and structure; however, polyphenolic phytoconstituents are the essential part of this composition that differ phytosomes from liposomes (Semalty et al. 2010). This makes them more stable than liposomes in biological fluids and harsh mediums like GIT. Phytosomes are one of the excellent carriers for plant extract as they have high capacity to load them via phospholipids interactions. Usually, they are manufactured from solid dispersion of plant extracts in phospholipids matrix. Phytosomes are manufactured using solvent evaporation, anti-solvent, and super-critical fluid methods (Gnananath et al. 2017). MERIVA® is a commercial product of curcumin based phytosomes developed for its improved oral delivery.

3.5.4 Metallic Nanoparticles

Metallic nanoparticles such as iron oxide, gold, silver, zinc, silica have been utilized in cancer therapy since long times. They are essentially stable with offering of diagnostics properties in many cases, easy to produce with narrow size ranges of nanoparticles, i.e. less than 100 nm easily. They can be converted into structure like nanocages and nanoshells as well (Sharma et al. 2018).

3.5.4.1 Iron Nanoparticles

Superparamagnetic iron oxide nanoparticles (SPIONs) are promising option for both simultaneous controlled delivery and diagnosis of cancers (Mahmoudi et al. 2011). They can be made site-directed by applying external magnetic field. SPIONs are safe as drug carrier and can be tailored as per the need of drug delivery. Curcumin loaded core-shell SPIONs have been utilized for its controlled delivery (Justin et al. 2018). A large number of processes of developing magnetic iron nanoparticles are available like microemulsion, sol–gel synthesis, sonochemical reactions, hydrothermal reactions, precursor hydrolysis and thermolysis, electrospray synthesis, flow injection synthesis, and chemical co-precipitation. The later one is one of the most common methods due to its simplicity and better control of the factors to generate good practical yield. However, it has less control over size distribution of the nanoparticles. SPIONS can be further modified using some polymer to make them surface engineered and also for better encapsulation of phytoconstituents (Dulińska-Litewka et al. 2019).

3.5.4.2 Gold Nanoparticles

Gold nanoparticles provide advantages of controlled drug delivery along with diagnosis option in chemoprevention. They are suitable drug delivery system offering desired essential features for anticancer nanotherapeutics. A comparatively high surface area for improved drug loading, easy surface functionalization required for intracellular uptake, control over size, morphology, and surface charge, easy surface modifications, and improved selectivity due to multivalent avidity. With these features, gold nanoparticles can accommodate higher number of ligands than other nanocarriers systems like liposome or polymeric nanoparticles, etc. The unique plasmonic properties of gold nanoparticles are useful for optically triggered control release of loaded phytoactives (Katti et al. 2009).

3.5.4.3 Silver Nanoparticles

Apart from nanosize, silver nanoparticles possess certain attractive features as drug carrier system like chemical stability, surface plasma resonance, high conductivity, and catalytic activity. Pharmacologically, they have been found active as antibacterial, wound healing property, and inhibition of VEGF induced angiogenesis. Both chemical and green methods of synthesis of silver nanoparticles are reported in literature. The structure and morphology of silver nanoparticles depend upon method adopted for synthesis. Silver nanoparticles synthesis is an easy procedure and provides better control over size, shape, and crystal formation (Abdel-Fattah and Ali 2018).

3.5.4.4 Silica Nanoparticles

Another class of inorganic nanoparticles based drug delivery system is mesoporous silica nanoparticles (MSNPs). These carriers offer high surface area along with specific pore volume, both thermal and chemical stability and surface functionalization opportunity. It is reported in the literature that MSNPs can be readily internalized by cells via endocytosis mechanism and these nanoparticles have high biodegradability. They are safe and biocompatible nanocarriers that can be utilized for chemoprevention using phytoconstituents loaded into it (Chaudhary et al. 2019).

3.5.5 Nanocapsules

Nanocapsules are core-shell type of nanosystem which provide protection to the loaded drug and also have capacity to target the drugs at various body parts. It provides controlled release of the phytoconstituents loaded into it. Plant extracts can be easily encapsulated in the core of the nanocapsules. Structurally, they are consisted of small droplets of a liquid core thoroughly packed by polymeric shell wall. This polymeric shell controls over the release of the loaded bioactives. By this way, nanocapsules increase the therapeutic efficacy, avoid degradation, and reducing unwanted toxicity as well. Nanocapsules provide excellent stability against microenvironment factors like light, enzyme, and pH, particularly encounter during oral administration (Murthy et al. 2018).

3.5.6 Hybrid Nanosystems

Nowadays, hybrid nanosystem composed of lipid–polymer, lipid–metal, and polymer–protein is being used to combine the benefits of both the biomaterials and avoid the drawbacks associated with individual material. Hybrid nanosystems are consisted of two different biomaterials structuring core and shell (Drakalska et al. 2014). In general, core is composed of metallic material which is coated with either lipid or polymers. In another case, lipid may be internal core and coated with polymers to provide it additional value addition or protection from harsh environments like GIT. Such types of systems orient with the good qualities of both the biomaterials and synergize the drug delivery capabilities of both together. A better control over the release of the loaded phytoconstituents could be achieved by protecting them as well.

3.6 Conclusion

Controlled delivery of chemopreventive agents requires tailoring of a desired carrier system. A variety of biomaterials are available with distinct unique properties and drug release characteristics. Further, the physicochemical properties of phytoconstituents are also crucial in selection of nanosystem for their control release. The route of administration, GRAS status of biomaterial, stability of active ingredients, scale-up possibility of the production methods, etc., are some more determining factors. Polymeric, lipid based (solid lipid or vesicular), metallic, and hybrid nanosystem can be chosen as per the need of the delivery required for chemoprevention.

References

Abdel-Fattah W, Ali GW (2018) On the anti-cancer activities of silver nanoparticles. J Appl Biotechnol Bioeng 5(1):43–46

Belcaro G, Cesarone MR, Dugall M, Pellegrini L, Ledda A, Grossi MG, Togni S, Appendino G (2010) Efficacy and safety of Meriva®, a curcumin-phosphatidylcholine complex, during extended administration in osteoarthritis patients. Altern Med Rev 15(4):337–344

Bharali DJ, Siddiqui IA, Adhami VM, Chamcheu JC, Aldahmash AM, Mukhtar H, Mousa SA (2011) Nanoparticle delivery of natural products in the prevention and treatment of cancers: current status and future prospects. Cancers 3(4):4024–4045

Bolger GT, Licollari A, Tan A, Greil R, Vcelar B, Greil-Ressler S, Weiss L, Schönlieb C, Magnes T, Radl B, Majeed M (2019) Pharmacokinetics of liposomal curcumin (Lipocurc™)

infusion: effect of co-medication in cancer patients and comparison with healthy individuals. Cancer Chemother Pharmacol 83(2):265–275

Chaudhary Z, Subramaniam S, Khan GM, Abeer MM, Qu Z, Janjua T, Kumeria T, Batra J, Popat A (2019) Encapsulation and controlled release of resveratrol within functionalized mesoporous silica nanoparticles for prostate cancer therapy. Front Bioeng Biotechnol 7:225

Davatgaran-Taghipour Y, Masoomzadeh S, Farzaei MH, Bahramsoltani R, Karimi-Soureh Z, Rahimi R, Abdollahi M (2017) Polyphenol nanoformulations for cancer therapy: experimental evidence and clinical perspective. Int J Nanomedicine 12:2689

Desai P, Thumma NJ, Wagh PR, Zhan S, Ann D, Wang J, Prabhu S (2020) Cancer chemoprevention using nanotechnology-based approaches. Front Pharmacol 11

Drakalska E, Momekova D, Manolova Y, Budurova D, Momekov G, Genova M, Antonov L, Lambov N, Rangelov S (2014) Hybrid liposomal PEGylated calix [4] arene systems as drug delivery platforms for curcumin. Int J Pharm 472(1-2):165–174

Dulińska-Litewka J, Łazarczyk A, Hałubiec P, Szafrański O, Karnas K, Karewicz A (2019) Superparamagnetic iron oxide nanoparticles—current and prospective medical applications. Materials 12(4):617

Elsabahy M, Wooley KL (2012) Design of polymeric nanoparticles for biomedical delivery applications. Chem Soc Rev 41(7):2545–2561

Elsheikh MA, Elnaggar YS, Otify DY, Abdallah OY (2018) Bioactive-chylomicrons for oral lymphatic targeting of berberine chloride: novel flow-blockage assay in tissue-based and caco-2 cell line models. Pharm Res 35(1):18

Ezhilarasi PN, Karthik P, Chhanwal N, Anandharamakrishnan C (2013) Nanoencapsulation techniques for food bioactive components: a review. Food Bioprocess Technol 6(3):628–647

Gnananath K, Nataraj KS, Rao BG (2017) Phospholipid complex technique for superior bioavailability of phytoconstituents. Adv Pharm Bull 7(1):35

Hua S, De Matos MB, Metselaar JM, Storm G (2018) Current trends and challenges in the clinical translation of nanoparticulate nanomedicines: pathways for translational development and commercialization. Front Pharmacol 9:790

Jain S, Jain V, Mahajan SC (2014) Lipid based vesicular drug delivery systems. Adv Pharm 2014

Justin C, Samrot AV, Sahithya CS, Bhavya KS, Saipriya C (2018) Preparation, characterization and utilization of coreshell super paramagnetic iron oxide nanoparticles for curcumin delivery. PLoS One 13(7):e0200440

Katdare A, Chaubal M (eds) (2006) Excipient development for pharmaceutical, biotechnology, and drug delivery systems. CRC Press

Katti K, Chanda N, Shukla R, Zambre A, Suibramanian T, Kulkarni RR, Kannan R, Katti KV (2009) Green nanotechnology from cumin phytochemicals: generation of biocompatible gold nanoparticles. Int J Green Nanotechnol Biomed 1(1):B39–B52

Khan T, Gurav P (2018) PhytoNanotechnology: enhancing delivery of plant based anti-cancer drugs. Front Pharmacol 8:1002

Lagoa R, Silva J, Rodrigues JR, Bishayee A (2020) Advances in phytochemical delivery systems for improved anticancer activity. Biotechnol Adv 38:107382

Mahmoudi M, Sahraian MA, Shokrgozar MA, Laurent S (2011) Superparamagnetic iron oxide nanoparticles: promises for diagnosis and treatment of multiple sclerosis. ACS Chem Neurosci 2 (3):118–140

Martínez-Ballesta M, Gil-Izquierdo Á, García-Viguera C, Domínguez-Perles R (2018) Nanoparticles and controlled delivery for bioactive compounds: outlining challenges for new "smart-foods" for health. Foods 7(5):72

Murthy KC, Monika P, Jayaprakasha GK, Patil BS (2018) Nanoencapsulation: an advanced nanotechnological approach to enhance the biological efficacy of curcumin. In: Advances in plant phenolics: from chemistry to human health. American Chemical Society, pp 383–405

Paliwal R, Rai S, Vaidya B, Khatri K, Goyal AK, Mishra N, Mehta A, Vyas SP (2009) Effect of lipid core material on characteristics of solid lipid nanoparticles designed for oral lymphatic delivery. Nanomedicine 5(2):184–191

Paliwal R, Paliwal SR, Agrawal GP, Vyas SP (2011a) Biomimetic solid lipid nanoparticles for oral bioavailability enhancement of low molecular weight heparin and its lipid conjugates: in vitro and in vivo evaluation. Mol Pharm 8(4):1314–1321

Paliwal SR, Paliwal R, Agrawal GP, Vyas SP (2011b) Liposomal nanomedicine for breast cancer therapy. Nanomedicine 6(6):1085–1100

Paliwal R, Paliwal SR, Agrawal GP, Vyas SP (2012a) Chitosan nanoconstructs for improved oral delivery of low molecular weight heparin: in vitro and in vivo evaluation. Int J Pharm 422 (1-2):179–184

Paliwal SR, Paliwal R, Agrawal GP, Vyas SP (2012b) Targeted breast cancer nanotherapeutics: options and opportunities with estrogen receptors. Crit Rev Ther Drug Carrier Syst 29(5)

Paliwal SR, Paliwal R, Pal HC, Saxena AK, Sharma PR, Gupta PN, Agrawal GP, Vyas SP (2012c) Estrogen-anchored pH-sensitive liposomes as nanomodule designed for site-specific delivery of doxorubicin in breast cancer therapy. Mol Pharm 9(1):176–186

Paliwal R, Babu RJ, Palakurthi S (2014) Nanomedicine scale-up technologies: feasibilities and challenges. AAPS PharmSciTech 15(6):1527–1534

Paliwal SR, Paliwal R, Agrawal GP, Vyas SP (2016) Hyaluronic acid modified pH-sensitive liposomes for targeted intracellular delivery of doxorubicin. J Liposome Res 26(4):276–287

Rai S, Paliwal R, Gupta PN, Khatri K, Goyal AK, Vaidya B, Vyas SP (2008) Solid lipid nanoparticles (SLNs) as a rising tool in drug delivery science: one step up in nanotechnology. Curr Nanosci 4(1):30–44

Rajalakshmi S, Vyawahare N, Pawar A, Mahaparale P, Chellampillai B (2018) Current development in novel drug delivery systems of bioactive molecule plumbagin. Artif Cells Nanomed Biotechnol 46(Suppl 1):209–218

Ranjan AP, Mukerjee A, Helson L, Gupta R, Vishwanatha JK (2013) Efficacy of liposomal curcumin in a human pancreatic tumor xenograft model: inhibition of tumor growth and angiogenesis. Anticancer Res 33(9):3603–3609

Semalty A, Semalty M, Rawat MSM, Franceschi F (2010) Supramolecular phospholipids–polyphenolics interactions: the PHYTOSOME® strategy to improve the bioavailability of phytochemicals. Fitoterapia 81(5):306–314

Sharma A, Goyal AK, Rath G (2018) Recent advances in metal nanoparticles in cancer therapy. J Drug Target 26(8):617–632

Singh M, Devi S, Rana VS, Mishra BB, Kumar J, Ahluwalia V (2019) Delivery of phytochemicals by liposome cargos: recent progress, challenges and opportunities. J Microencapsul 36 (3):215–235

Subramanian AP, Jaganathan SK, Manikandan A, Pandiaraj KN, Gomathi N, Supriyanto E (2016) Recent trends in nano-based drug delivery systems for efficient delivery of phytochemicals in chemotherapy. RSC Adv 6(54):48294–48314

Nanochemoprevention

4

Abstract

This chapter deals with the pharmaceutical approaches of translating the chemopreventive phytochemicals into a useful nanochemopreventive product. Summary of the recent reports of nanochemoprevention based on promising nanocarriers like polymeric nanoparticles, lipid nanoparticles, polysaccharide based nanoparticles, selenium nanoparticles, gold nanoparticles, and liposomes have been described for the readers. The advantages of these nanocarriers in delivery and translation of chemopreventive agent as a possible commercial product are also elaborated with classic examples.

Keywords

Nanotechnology · Drug delivery · Chemoprevention · Phytochemicals · Anticancer herbal drug · Nanomedicine

4.1 Nanochemoprevention

Cancer chemoprevention through phytochemicals has been showed promising result in cancer management, however, the clinical translation has received limited success due to low bioavailability, high dose, and stability issues related to these natural compounds (Siddiqui et al. 2008). Therefore, to exploit full potential of chemopreventive agents some novel techniques are required that alleviate these limitations and enhance the therapeutic potential. Nanotechnology driven solutions for biomedical applications have created multiple dogmas to treat life threatening disease like cancer. Nanoparticles loaded with phytochemicals known for cancer chemoprevention may selectively target to the cancer cells and minimize pharmaceutical problems like low bioavailability, poor aqueous solubility, physiological stability issues of individual molecules and poor pharmacokinetics after in vivo

R. Paliwal, S. R. Paliwal, *Advances in Nanochemoprevention*,
https://doi.org/10.1007/978-981-15-9692-6_4

administration (Paliwal et al. 2014). Nano-phytochemicals have several benefits like biocompatibility, biodegradability, physiologic stability, specificity, enhanced absorption properties, less degradability, reduced toxicity along with reduced dose requirement and lower cost of treatment (Yadav et al. 2020). Table 4.1 lists biopharmaceutical limitations of natural chemopreventive agents that can be overcome by nanocarriers and possible benefit thereafter. In addition, a combination of chemopreventive agent with cytotoxic drugs can be achieved in one system to make them efficient in cancer cell killing via various alternative mechanisms. Being nanosize carrier, they can passively target the cancer bioenvironment through enhanced permeability retention (EPR) effect.

Phytochemicals with chemopreventive activity are no more exception similar to many other drugs and have poor oral bioavailability problems due to their poor water solubility and limited absorption through GIT after oral administration (Van Duynhoven et al. 2011). Increasing bioavailability and decreasing administered dose using nanoparticles may be a solution for this constraint especially when long-term treatment is required as in case of cancer therapy (Fig. 4.1). Nanoparticles can be developed from FDA approved synthetic polymers like PLGA or natural polymers or lipids to address above-mentioned issues (Shirode et al. 2015) (Fig. 4.2). However, nanoencapsulation of such molecules not only provide stability to these molecules but also enhance their cellular uptake and better cytotoxicity resulting in poor cancer cell growth (Fig. 4.3). Table 4.2 summarizes recent studies conducted with chemopreventive agents loaded nanoparticles. In the following sections, we described recent reports highlighting emerging role of nanoparticles in chemoprevention.

4.2 Polymeric Nanoparticles

Polymeric nanoparticles present the opportunities to encapsulate poorly soluble drugs, protect them and modify their blood circulation (Mishra et al. 2008). Chun and colleagues reported curcumin encapsulated in polymeric nanoparticles composed of isopropylacrylamide, vinylpyrrolidone, and acrylic acid (Chun et al. 2012). They observed a comparable chemopreventive effect of such formulations even when administered at 20-fold lower doses than oral curcumin in a carcinogen-induced mammary tumor. A significantly enhanced oral bioavailability of curcumin was achieved after incorporation into nanoparticles. Another example is use of a biocompatible blend of poly(epsilon-caprolactone) (PCL) and poly(d,l-lactic-co-glycolic acid)-poly(ethylene glycol) conjugate (PLGA–PEG–COOH) for trans-resveratrol (RSV) delivery (Sanna et al. 2013a, b). Particles were precipitated to load the drug into polymers. Such nanoprecipitated formulation referred as nano-RSV has shown significantly improved cytotoxicity than free RSV in different cell lines and considerably shown a consistent sensitivity towards both the androgen-independent and hormone-sensitive cells. PLGA nanoparticles of tea polyphenols the aflavin and EGCG have been developed recently to test for their protective effect against chemically induced DNA damage. Authors found that TF or EGCG loaded

Table 4.1 List of biopharmaceutical limitations of natural chemopreventive agents that can be overcome by nanocarriers and possible benefit thereafter

Natural compound	Limitation	Reported suitable nanocarrier	Possible benefits	Reference
Curcumin	• Poor aqueous solubility, low oral bioavailability (high dose approx. 8000 mg/day free curcumin orally is required in order to achieve detectable systemic levels) • Inefficient systemic delivery • Rapid metabolism in the intestines and liver	Encapsulated/entrapped/emulsified, or self-assembled in a carrier such as micellar aggregates, liposomes, biocompatible polymers like alginate, chitosan, and Pluronic, a PEGylated conjugate, PLGA–PEG NPs/PLGA-NPs, as nanocrystal solid dispersion, amorphous solid dispersion, or nanoemulsion of poly (oxyethylene) cholesteryl ether (PEGChol)	• Nanocarrier that can enable parenteral administration of curcumin in an aqueous phase will significantly increase the clinical applicability • Curcumin NPs exhibit higher half-lives and low serum clearance and enhanced cellular uptake and bioactivity • Significantly increased bioavailability, Cellular intake, antiproliferative, and anti cell-growth properties	Thangapazham et al. (2008), Hsieh (2001), Bisht et al. (2007) and Sahu et al. (2008)
EGCG	• High dose • Low bioavailability • Short half-lives • Oxidation labile • Inefficient systemic delivery	• Chitosan nanoparticles (CNs) • Poly(lactide-co-glycolide) nanoparticles (PLGA-NPs)	• To make a stable formulation in the acidic environment of the stomach • Steady and sustained release of EGCG • To increase retention time (mucoadhesive properties so they also adhere to the gastrointestinal tract) • PLGA-NPs have approx. 30-fold dose-advantage than bulk drug	Srivastava et al. (2013) and Siddiqui et al. (2014)

(continued)

Table 4.1 (continued)

Natural compound	Limitation	Reported suitable nanocarrier	Possible benefits	Reference
Naringenin	• Low water solubility • Poor bioavailability	β-cyclodextrin inclusion complex	• Increased water solubility • High thermal stability	Yang et al. (2013)
Resveratrol	• Extremely photosensitive compound with low chemical stability • Poor solubility, inefficient systemic delivery • Low bioavailability	Cationic chitosan (CS)- and anionic alginate (Alg)-coated poly(d, l-lactide-*co*-glycolide) nanoparticles (NPs)	• Improved drug loading • Effective controlled release • Protection against light exposure degradation	Sanna et al. (2012)
Kaempferol	• Inefficient systemic delivery • Limited bioavailability	• Poly(d,l-lactide-*co*-glycolide) nanoparticles	• Increased bioavailability	Luo et al. (2012)
Genistein	• Poor solubility • Low serum levels after oral administration	• Self-nanoemulsified systems • Super paramagnetic systems • Chitosan microspheres • Eudragit nanoparticles	• Increased dissolution and bioavailability	Si et al. (2010) and Tang et al. (2011)
Caffeine	• Hydrophilic drug • Poor transdermal delivery	• Nanoemulsions	• Improved transdermal permeation	Shakeel and Ramadan (2010)
Quercetin	• Low aqueous solubility • Instability in physiological medium • Poor bioavailability • Poor permeability • Extensive first pass metabolism	• Poly-d,l-lactide (PLA) nanoparticles • Liposomes • Chitosan nanoparticles • Solid lipid nanoparticles	• Improve aqueous solubility and stability	Kumari et al. (2010), Priprem et al. (2008), Zhang et al. (2008) and Li et al. (2009)

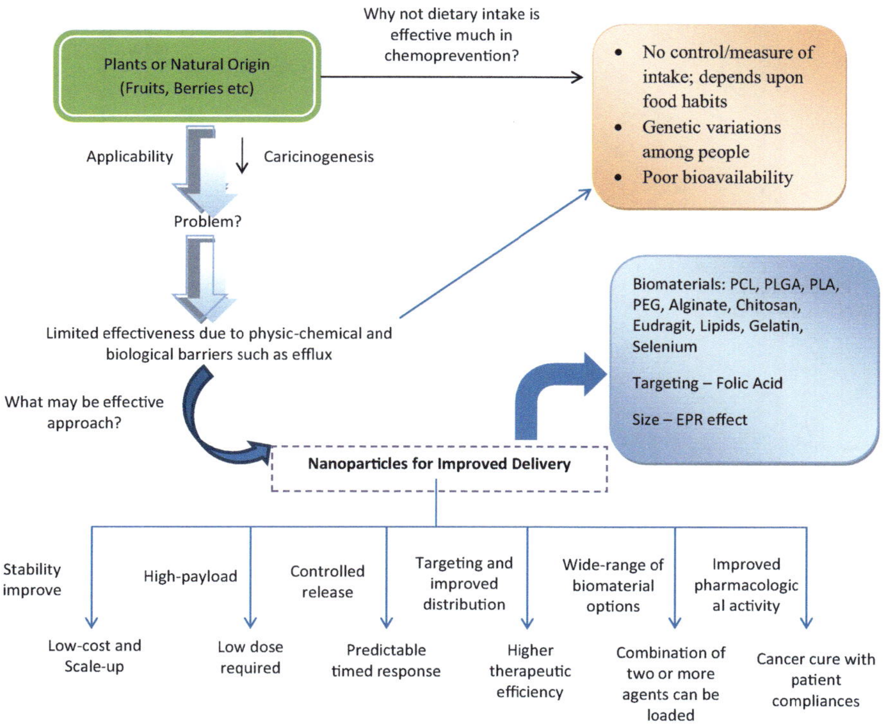

Fig. 4.1 Schematic presentation of wide-range applications of nanoparticles for improved delivery of chemopreventive agents through enlisted biomaterials and selective targeting approaches

PLGA-NPs have significant potential for induction of DNA repair genes and suppression of DNA damage responsive genes as compared with respective bulk TF or EGCG doses. Even when used in very less concentration, TF or EGCG loaded PLGA-NPs have shown superior ability to prevent DMBA-induced DNA damage than free drug.

In a similar study conducted by another research group, the effects of TFs and EGCG were compared in the bulk form and in the polymer (poly[lactic-co-glycolic acid])-based NP form, in oxaliplatin- and satraplatin-treated lymphocytes as surrogate cells from colorectal cancer patients and healthy volunteers (Alotaibi et al. 2013). It was claimed that both the compounds in the bulk form produced concentration-dependent reductions in DNA damage in oxaliplatin- and satraplatin-treated lymphocytes. Whereas NP form of both TFs and EGCG although initially caused a reduction, but later they produced a concentration-dependent increase in DNA damage in the lymphocytes. It was concluded that TFs and EGCG act as both antioxidants and pro-oxidants, however, their activity depends on the form in which they are being tested. Khan et al. (2014) developed chitosan based nanoparticles loaded with EGCG and tested their potential for reduction in

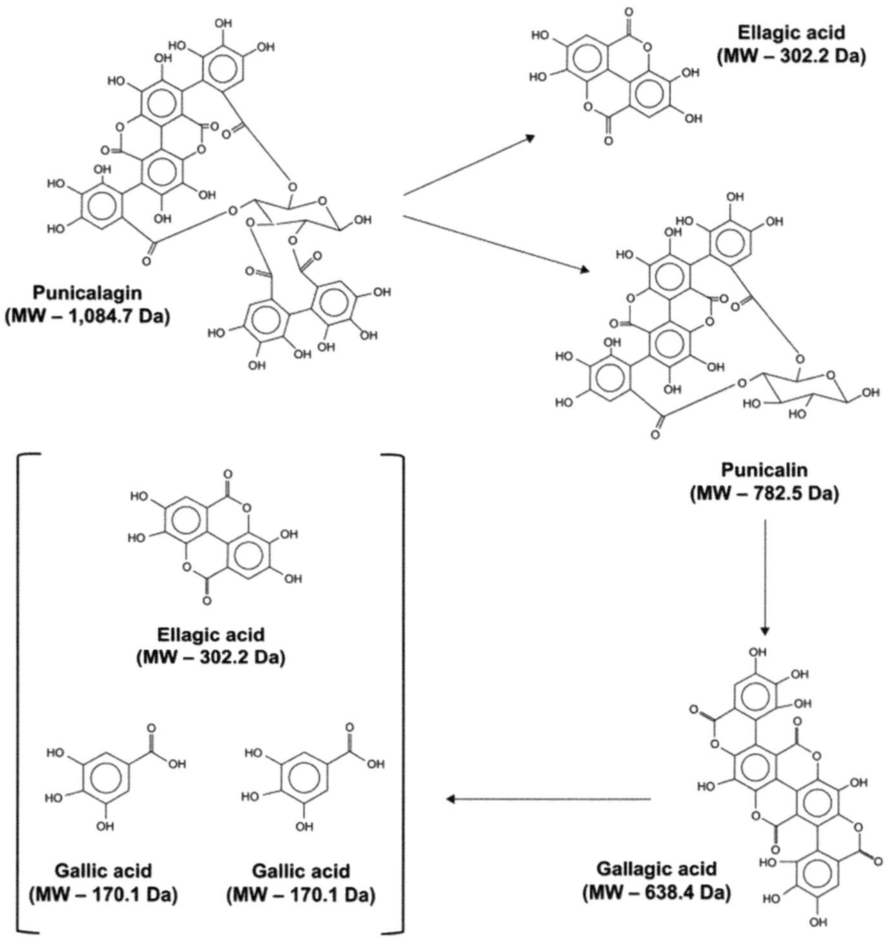

Fig. 4.2 Representative example of instability of natural chemopreventive agents; Hydrolysis of punicalagin, a major polyphenol in pomegranates, which is not absorbed in its intact form but is hydrolyzed to ellagic acid (EA) moieties and rapidly metabolized into short-lived metabolites of EA (Adopted from Shirode et al. 2015)

tumor growth among athymic nude mice (Khan et al. 2014). Authors found the superior efficacy of nano-complexed EGCG over plain drug.

4.3 Lipid Nanoparticles

Lipid nanoparticles offer an alternative option to polymeric nanoparticles for controlled and targeted delivery of chemopreventive agents with poor aqueous solubility (Vyas et al. 2008). Solid lipid nanoparticles (SLNs) are colloidal carriers of small

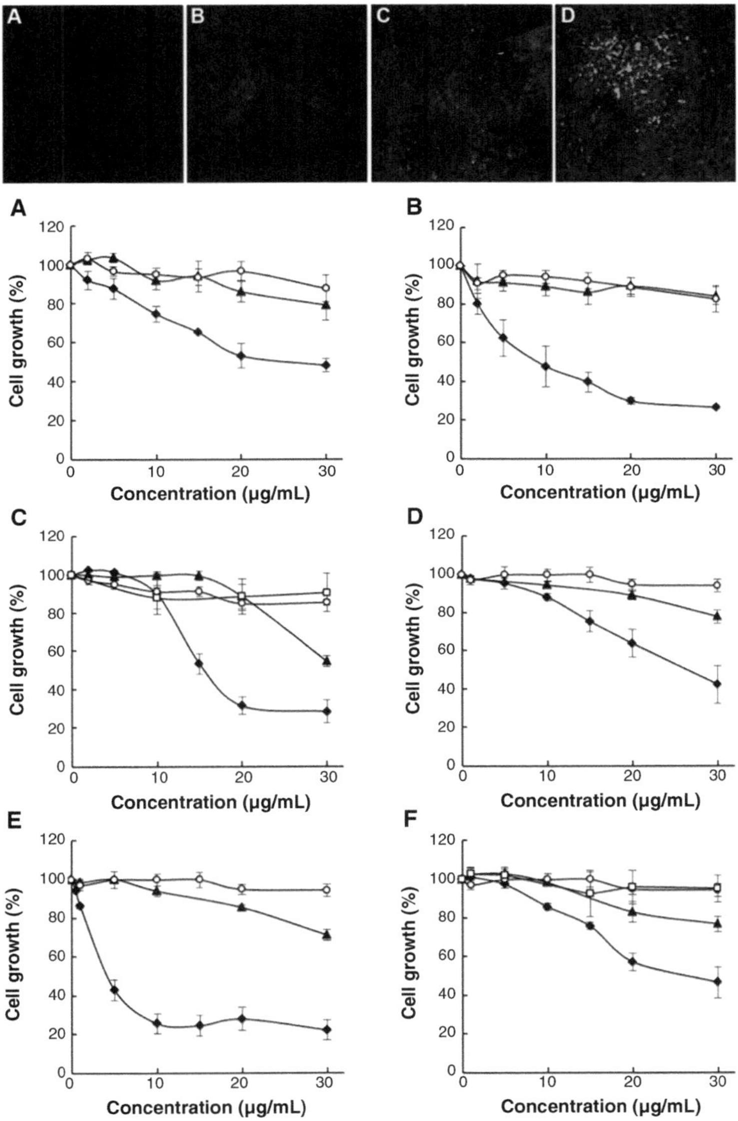

Fig. 4.3 Uptake and proven superior efficacy of nanoparticles developed for delivery of chemopreventive bioactives [**I**] Intracellular uptake of NPs over a 24-h time course, MCF-7 cells were incubated with Alexa Fluor-488-labeled PLGA–PEG NPs for 15 min (**a**), 2 h (**b**), 6 h (**c**), and 24 h (**d**)); [**II**] The effect of pomegranate nanoprototypes on MCF-7 and Hs578T breast cancer cell growth. MCF-7 cells were treated with PE-NPs (**a**), PU-NPs (**b**), and EA-NPs (**c**) versus their respective free counterparts. Hs578T cells were treated with PE-NPs (**d**), PU-NPs (**e**), and EA-NPs (**f**) versus their respective free counterparts. Solid diamonds show pomegranate phytochemical loaded NPs, solid triangles show respective free counterparts, open circles depict void NPs, and open rectangles show NaOH solvent control for EA. (Adopted from Shirode et al. 2015)

Table 4.2 List of recent studies conducted with chemopreventive agents loaded nanoparticles

Carrier/Material	Chemopreventive agent(s)	Cancer studied	In vitro/In vivo study	Significant outcomes	Reference
PLGA, PEO–PPO–PEO, Glycol chitosan, PLGA–PEI, and PAMAM dendrimer	Kaempferol	Ovarian cancer	IOSE397 cells, A2780/CP70, and OVCAR-3	Both PEO–PPO–PEO and PLGA nanoparticle formulations had superior effects compared with kaempferol alone in reducing cancer cell viability	Luo et al. (2012)
Solid lipid nanoparticles (SLN)	Aspirin, Curcumin, and Sulforaphane (ACS)	Pancreatic cancer	Syrian golden hamster model	Oral, low-dose, nanotechnology-based combinatorial regimen developed for the long-term chemoprevention of pancreatic cancer	Grandhi et al. (2013)
Poly(epsilon-caprolactone) and poly(D,L-lactic-co-glycolic acid)-poly(ethylene glycol) blend	Resveratrol	Prostate cancer	PCa cell lines	Potential use of developed nanoprototypes for the controlled delivery of bioactive RSV for PCa chemoprevention/chemotherapy	Sanna et al. (2013a, b)
Eudragit® E	Naringenin	Experimental oral carcinogenesis	DMBA-induced experimental oral carcinogenesis	Oral administration of NARNPs (50 mg NAR/kg body weight/day) to DMBA-treated animals completely prevented the tumor formation as compared to the free NAR	Sulfikkarali et al. (2013)
Chitosan	EGCG	Melanoma	Mel 928 tumor xenograft mice model	Nano-EGCG treated cells showed marked induction of apoptosis and cell cycle inhibition along with the growth of Mel 928 tumor xenograft	Siddiqui et al. (2014)
Water-soluble polymer	Luteolin	Lung cancer and head and neck cancer	Tumor xenograft mouse model	Nano-Luteolin has a significant inhibitory effect on the tumor growth of squamous cell carcinoma of head and neck cancer in comparison to luteolin	Majumdar et al. (2014)

Chitosan-coated solid lipid nanoparticles (c-SLN)	Ferulic acid (FA), an antioxidant, combined with aspirin (ASP)	Pancreatic cancer chemoprevention	MIA PaCa-2 and Panc-1	When encapsulated within c-SLNs, a 5- and 40-fold decrease in dose of FA (40 μM) and ASP (25 μM) was observed which was significant. Oral administration of combinations of 75 and 25 mg/kg of FA and ASP c-SLNs to MIA PaCa-2 pancreatic tumor xenograft mice model suppressed the growth of the tumor by 45% compared to control	Thakkar et al. (2015)
PLGA	Curcumin	Prostate cancer	Prostate cancer xenograft mice model	PLGA-CUR NPs can significantly accumulate and exhibit superior anticancer activity in prostate cancer	Yallapu et al. (2014)
Polycaprolactone, TPGS	Resveratrol	Colon	HT29 cell lines	Resveratrol loaded lipid cored nanocapsules showed higher efficacy with approximate 36% cell apoptosis	Feng et al. (2017a, b)
PLGA	Resveratrol	Prostate	LNCaP cells	Nanoparticles significantly decreased the cell viability mediated by apoptosis	Nassir et al. (2018)
Lipid nanoparticles	Curcumin	Breast cancer	MCF10A, MCF7, and MDA-MB-231 cell lines	The developed formulation showed antioxidant and antitumor effects and radiosensitizing effect	Minafra et al. (2019)
Human serum albumin nanoparticles (HSA-CUR)	Curcumin	Breast cancer	MCF7 and SK-BR3	HSA-CUR nanoparticles increased water solubility and physiologic stability. Nanoparticles revealed high anticancer potential as compared to free drug	Matloubi and Hassan (2020)

size, biocompatible in nature, protect content against chemical degradation and their surface can be easily modified. Being lipidic in nature SLNs delivered a major fraction of administered drug through the lymphatic system, thus avoid first pass metabolism (Paliwal et al. 2011a, b).

SLN formulations loaded with cytotoxic agents have shown better efficacy and cytotoxicity in comparison to free drug itself (Paliwal et al. 2009; Chaudhary et al. 2011). Due to physicochemical characteristics and controlled release kinetics properties, SLN would be a carrier of choice for the delivery of lipophilic chemopreventive agents such as curcumin, flavonoids, etc. However, they suffer with the problem of low entrapment efficiency and storage stability. Grandhi et al. (2013) developed a lipid based nanoformulation for oral, long-dose, and long-term chemoprevention against pancreatic cancer (Grandhi et al. 2013). Aspirin, curcumin, and sulforaphane (ACS) combinations were tested in three different doses (low, medium, and high). It was observed that unmodified ACS combinations exhibited reduction in tumor incidence by 18%, 50%, and 68.7%, respectively; whereas the modified nanoencapsulated ACS regimens reduced tumor incidence by 33%, 67%, and 75%, respectively, at 10 times lower dose of free drug combinations. Conclusively, using lipid based nanocarrier, dose of chemopreventive agent can be reduced without compromising tumor regression as it was reported in this case. Minafra et al. (2019) prepared solid lipid nanoparticles loaded with curcumin and evaluated radiosensitizing potential on breast cancer cell lines. They used transcriptomic and metabolomic strategies to study antitumor and antioxidant efficacy and suggested that the developed formulation can be successfully used for chemopreventive action.

4.4 Polysaccharide Nanoparticles

Polysaccharides such as chitosan, alginate, gum arabic, and maltodextrin, etc., have been used as either carriers or surface modifiers for improved delivery of chemopreventive agents to cancer cells. Rocha et al. (2011) incorporated polyphenol EGCG in the gum arabic and maltodextrin matrix with about 85% incorporation efficiency (Rocha et al. 2011). It was found that EGCG retained its biological activity after encapsulation and reduced the cell viability by inducing apoptosis in Du145 prostate cancer cells. Its inhibitory effect on cell proliferation was 10–20% at lower concentrations compared with free EGCG. Polysaccharides such as chitosan and alginates have also been used to coat nano/micro-particles of other polymers in order to provide better pharmaceutical perspective. In such a study, cationic chitosan and anionic alginate coated PLGA nanoparticles loaded with resveratrol have been developed and characterized for in vitro performance (Sanna et al. 2012). Authors found that nanocarriers showed a biphasic release pattern with more effective controlled release rates for NPs coated with higher polyelectrolyte concentrations. The trans-cis photoisomerization reaction, a light exposure stability problem with resveratrol was minimized when drug was loaded in the developed system.

4.5 Selenium Nanoparticles

Selenium (Se) is known for its cytotoxic potential when used in higher concentration than its dietary amount needed. Though, it has low therapeutic window in between the two. Nanoform of Se (Nano-Se) possesses lower toxicity than Se alone but with therapeutic activity at this level to increase the activities of selenoenzymes and phase II enzymes. In the nanoform, Se itself is found chemopreventive with less toxicity concern than Se. Nano-Se has been used along with other chemotherapeutic agents to combat the cancer. In a recent study, Nano-Se along with irinotecan was tested for its chemopreventive potential and synergistic enhancement of the antitumor treatment effect in vitro on HCT-8 tumor cells and in vivo on xenografted tumor both (Gao et al. 2014). Authors highlighted that the beneficial effects of Nano-Se for tumor therapy were associated to selective regulation of Nrf2-ARE (antioxidant responsive elements) pathway in tumor tissues and normal tissues. Further, Se-drug complex can be targeted to selectively to cancerous cells using ligand–receptor interactions. Recently, folate conjugated selenium–rubyrin loaded nanoparticles with an acidic pH-activatable targeted photosensitizer have been reported in the literature. Such carriers specifically recognize cancer cells via the FA–FA receptor binding followed by selective internalization by the cancer cells to enter lysosomes and ultimately activated to produce reactive oxygen. Mechanistically, these carriers killed the cells through lysosome-associated pathway. When tested in vivo in tumor-bearing mice, i.v. administration of FA conjugated Se–rubyrin nanoparticles resulted into evident tumor elimination after NIR exposure. In the future, few other targeting agents need to be utilized to prove superior efficacy of targeted selenium as multifunctional chemopreventive agent.

4.6 Gold Nanoparticles

Gold nanoparticles due to their unique applications in diagnostic and cancer therapy serve as indispensible tool of research when it comes to molecular targeted nanomedicine. They are one of the preferred nanocarriers for chemoprevention. During synthesis of gold nanoparticle from gold solution, phytochemical extracts can be easily loaded. Such prepared system may be useful for cell detection as well as killing cancerous cells. By means of this approach, Mukherjee et al. (2015) developed gold-conjugated green tea nanoparticles. These carriers were aimed to enhance antitumor activity of green tea *vis a vis* hepatoprotection (Mukherjee et al. 2015). Green tea polyphenols reduced gold into gold nanoparticles in just one step. An obvious advantage of this green approach of NPs production is complete avoidance of any organic solvent which make it safe for internal human use. Such NPs have shown excellent selective toxicity towards Ehrlich's Ascites Carcinoma and MCF-7 cells. On normal primary mouse hepatocytes, these particles were absolutely safe and have shown no lethality when tested. At present, this research area is in infancy and hence only little reports are available for utilization of gold nanoparticles in chemoprevention. However, it will be an interesting area to be investigated in the future because of theragnostic applications of Gold nanoparticles in addition.

4.7 Liposomes

Liposomes have been widely explored as nanocarriers for delivery of phytochemicals because of biocompatibility, enhanced bioavailability, and their surface can be easily modified for targeted delivery for cancer treatment (Paliwal et al. 2016). Both hydrophilic and hydrophobic agents can be loaded in liposomes and are promising carrier for delivery of natural components. Feng et al. (2017a, b) reviewed the application of liposomes for the curcumin delivery and recommended that the encapsulation within the liposomes enhances bioavailability, stability, and anticancer property of curcumin while reducing the dose and toxicity (Feng et al. 2017a, b). Wu et al. (2018) developed luteolin loaded liposomes to enhance antitumor efficacy and studied the mechanism of their antitumor action on colorectal carcinoma. The result showed that the liposomal luteolin significantly reduced the growth of CT26 cells and higher antitumor effect as compared to free drug, suggesting the potential applicability of liposomal systems to increase solubility, stability, and bioavailability of chemopreventive agents (Wu et al. 2018). In a study, TPGS coated resveratrol loaded liposomes were developed for passive targeting to brain cancer. TPGS coated Res-liposomes showed longer circulation time and improved pharmacokinetic parameters and they suggested that Res-liposomes and TPGS-Res-liposomes were promising carrier for brain cancer treatment (Vijayakumar et al. 2016). Sailor et al. (2015) prepared berberine encapsulated liposomes and compared the release profile with suspension. They reported that drug was released completely within 10 h from suspension while only 70% release was observed from liposomes in 24 h, thus they concluded that the liposomal encapsulation controlled the release of drug (Sailor et al. 2015).

4.8 Conclusion

Nanochemoprevention requires a construction of a suitable nanocarrier loaded with phytoconstituents for its improved bioavailability at the target cancer site. This could be achieved due to the passive delivery of nanoparticles as a result of EPR effect. Among the several available options, polymeric nanoparticles provide ease of production and scale-up; however, render the regulatory issues of the process or the biomaterials. Lipid based nanoparticles are the alternative carriers as they are composed of biocompatible lipids; however, the clinical success depends upon the commercialization possibility of the process and quantity of the surfactant used. Liposomes are one of the successful nanomedicine products and develop hopes for the nanochemopreventive products as well. Selenium and Gold nanoparticles are proven efficacious systems for chemopreventive and diagnostics options. Almost all, clinically effective phytochemicals have been loaded and tested for their chemopreventive and cytotoxic action using these nanosystems.

References

Alotaibi A, Bhatnagar P, Najafzadeh M, Gupta KC, Anderson D (2013) Tea phenols in bulk and nanoparticle form modify DNA damage in human lymphocytes from colon cancer patients and healthy individuals treated in vitro with platinum-based chemotherapeutic drugs. Nanomedicine 8(3):389–401

Bisht S, Feldmann G, Soni S, Ravi R, Karikar C, Maitra A, Maitra A (2007) Polymeric nanoparticle-encapsulated curcumin ("nanocurcumin"): a novel strategy for human cancer therapy. J Nanobiotechnol 5(1):3

Chaudhary A, Sutaria D, Huang Y, Wang J, Prabhu S (2011) Chemoprevention of colon cancer in a rat carcinogenesis model using a novel nanotechnology-based combined treatment system. Cancer Prev Res 4(10):1655–1664

Chun YS, Bisht S, Chenna V, Pramanik D, Yoshida T, Hong SM, de Wilde RF, Zhang Z, Huso DL, Zhao M, Rudek MA (2012) Intraductal administration of a polymeric nanoparticle formulation of curcumin (NanoCurc) significantly attenuates incidence of mammary tumors in a rodent chemical carcinogenesis model: implications for breast cancer chemoprevention in at-risk populations. Carcinogenesis 33(11):2242–2249

Feng T, Wei Y, Lee RJ, Zhao L (2017a) Liposomal curcumin and its application in cancer. Int J Nanomed 12:6027

Feng M, Zhong LX, Zhan ZY, Huang ZH, Xiong JP (2017b) Enhanced antitumor efficacy of resveratrol-loaded nanocapsules in colon cancer cells: physicochemical and biological characterization. Eur Rev Med Pharmacol Sci 21(2):375–382

Gao F, Yuan Q, Gao L, Cai P, Zhu H, Liu R, Wang Y, Wei Y, Huang G, Liang J, Gao X (2014) Cytotoxicity and therapeutic effect of irinotecan combined with selenium nanoparticles. Biomaterials 35(31):8854–8866

Grandhi BK, Thakkar A, Wang J, Prabhu S (2013) A novel combinatorial nanotechnology-based oral chemopreventive regimen demonstrates significant suppression of pancreatic cancer neoplastic lesions. Cancer Prev Res 6(10):1015–1025

Hsieh CY (2001) Phase I clinical trial of curcumin, a chemopreventive agent, in patients with high-risk or pre-malignant lesions. Anticancer Res 21(2895):e2900

Khan N, Bharali DJ, Adhami VM, Siddiqui IA, Cui H, Shabana SM, Mousa SA, Mukhtar H (2014) Oral administration of naturally occurring chitosan-based nanoformulated green tea polyphenol EGCG effectively inhibits prostate cancer cell growth in a xenograft model. Carcinogenesis 35 (2):415–423

Kumari A, Yadav SK, Pakade YB, Singh B, Yadav SC (2010) Development of biodegradable nanoparticles for delivery of quercetin. Colloids Surf B: Biointerfaces 80(2):184–192

Li H, Zhao X, Ma Y, Zhai G, Li L, Lou H (2009) Enhancement of gastrointestinal absorption of quercetin by solid lipid nanoparticles. J Control Release 133(3):238–244

Luo H, Jiang B, Li B, Li Z, Jiang BH, Chen YC (2012) Kaempferol nanoparticles achieve strong and selective inhibition of ovarian cancer cell viability. Int J Nanomed 7:3951

Majumdar D, Jung KH, Zhang H, Nannapaneni S, Wang X, Amin AR, Chen Z, Shin DM (2014) Luteolin nanoparticle in chemoprevention: in vitro and in vivo anticancer activity. Cancer Prev Res 7(1):65–73

Matloubi Z, Hassan Z (2020) HSA-curcumin nanoparticles: a promising substitution for curcumin as a cancer chemoprevention and therapy. DARU 28(1):209–219

Minafra L, Porcino N, Bravatà V, Gaglio D, Bonanomi M, Amore E, Cammarata FP, Russo G, Militello C, Savoca G, Baglio M (2019) Radiosensitizing effect of curcumin-loaded lipid nanoparticles in breast cancer cells. Sci Rep 9(1):1–16

Mishra N, Goyal AK, Khatri K, Vaidya B, Paliwal R, Rai S, Mehta A, Tiwari S, Vyas S, Vyas SP (2008) Biodegradable polymer based particulate carrier (s) for the delivery of proteins and peptides. Anti-Inflammatory Anti-Allergy Agents Med Chem (Formerly Current Medicinal Chemistry-Anti-Inflammatory and Anti-Allergy Agents) 7(4):240–251

Mukherjee S, Ghosh S, Das DK, Chakraborty P, Choudhury S, Gupta P, Adhikary A, Dey S, Chattopadhyay S (2015) Gold-conjugated green tea nanoparticles for enhanced anti-tumor activities and hepatoprotection—synthesis, characterization and in vitro evaluation. J Nutr Biochem 26(11):1283–1297

Nassir AM, Shahzad N, Ibrahim IA, Ahmad I, Md S, Ain MR (2018) Resveratrol-loaded PLGA nanoparticles mediated programmed cell death in prostate cancer cells. Saudi Pharmaceutical J 26(6):876–885

Paliwal R, Rai S, Vaidya B, Khatri K, Goyal AK, Mishra N, Mehta A, Vyas SP (2009) Effect of lipid core material on characteristics of solid lipid nanoparticles designed for oral lymphatic delivery. Nanomedicine 5(2):184–191

Paliwal R, Paliwal SR, Agrawal GP, Vyas SP (2011a) Biomimetic solid lipid nanoparticles for oral bioavailability enhancement of low molecular weight heparin and its lipid conjugates: in vitro and in vivo evaluation. Mol Pharm 8(4):1314–1321

Paliwal R, Rai S, Vyas SP (2011b) Lipid drug conjugate (LDC) nanoparticles as autolymphotrophs for oral delivery of methotrexate. J Biomed Nanotechnol 7(1):130–131

Paliwal R, Babu RJ, Palakurthi S (2014) Nanomedicine scale-up technologies: feasibilities and challenges. AAPS PharmSciTech 15(6):1527–1534

Paliwal SR, Paliwal R, Agrawal GP, Vyas SP (2016) Hyaluronic acid modified pH-sensitive liposomes for targeted intracellular delivery of doxorubicin. J Liposome Res 26(4):276–287

Priprem A, Watanatorn J, Sutthiparinyanont S, Phachonpai W, Muchimapura S (2008) Anxiety and cognitive effects of quercetin liposomes in rats. Nanomedicine 4(1):70–78

Rocha S, Generalov R, Pereira MDC, Peres I, Juzenas P, Coelho MA (2011) Epigallocatechin gallate-loaded polysaccharide nanoparticles for prostate cancer chemoprevention. Nanomedicine 6(1):79–87

Sahu A, Kasoju N, Bora U (2008) Fluorescence study of the curcumin– casein micelle complexation and its application as a drug nanocarrier to cancer cells. Biomacromolecules 9 (10):2905–2912

Sailor G, Seth AK, Parmar G, Chauhan S, Javia A (2015) Formulation and in vitro evaluation of berberine containing liposome optimized by 32 full factorial designs. J Appl Pharm Sci 5 (7):023–028

Sanna V, Roggio AM, Siliani S, Piccinini M, Marceddu S, Mariani A, Sechi M (2012) Development of novel cationic chitosan-and anionic alginate–coated poly (d, l-lactide-co-glycolide) nanoparticles for controlled release and light protection of resveratrol. Int J Nanomed 7:5501

Sanna V, Siddiqui IA, Sechi M, Mukhtar H (2013a) Nanoformulation of natural products for prevention and therapy of prostate cancer. Cancer Lett 334(1):142–151

Sanna V, Siddiqui IA, Sechi M, Mukhtar H (2013b) Resveratrol-loaded nanoparticles based on poly (epsilon-caprolactone) and poly (d, l-lactic-co-glycolic acid)–poly (ethylene glycol) blend for prostate cancer treatment. Mol Pharm 10(10):3871–3881

Shakeel F, Ramadan W (2010) Transdermal delivery of anticancer drug caffeine from water-in-oil nanoemulsions. Colloids Surf B: Biointerfaces 75(1):356–362

Shirode AB, Bharali DJ, Nallanthighal S, Coon JK, Mousa SA, Reliene R (2015) Nanoencapsulation of pomegranate bioactive compounds for breast cancer chemoprevention. Int J Nanomed 10:475

Si HY, Li DP, Wang TM, Zhang HL, Ren FY, Xu ZG, Zhao YY (2010) Improving the anti-tumor effect of genistein with a biocompatible superparamagnetic drug delivery system. J Nanosci Nanotechnol 10(4):2325–2331

Siddiqui IA, Afaq F, Adhami VM, Mukhtar H (2008) Prevention of prostate cancer through custom tailoring of chemopreventive regimen. Chem Biol Interact 171(2):122–132

Siddiqui IA, Bharali DJ, Nihal M, Adhami VM, Khan N, Chamcheu JC, Khan MI, Shabana S, Mousa SA, Mukhtar H (2014) Excellent anti-proliferative and pro-apoptotic effects of (−)-epigallocatechin-3-gallate encapsulated in chitosan nanoparticles on human melanoma cell growth both in vitro and in vivo. Nanomedicine 10(8):1619–1626

Srivastava AK, Bhatnagar P, Singh M, Mishra S, Kumar P, Shukla Y, Gupta KC (2013) Synthesis of PLGA nanoparticles of tea polyphenols and their strong in vivo protective effect against chemically induced DNA damage. Int J Nanomedicine 8:1451

Sulfikkarali N, Krishnakumar N, Manoharan S, Nirmal RM (2013) Chemopreventive efficacy of naringenin-loaded nanoparticles in 7, 12-dimethylbenz (a) anthracene induced experimental oral carcinogenesis. Pathol Oncol Res 19(2):287–296

Tang J, Xu N, Ji H, Liu H, Wang Z, Wu L (2011) Eudragit nanoparticles containing genistein: formulation, development, and bioavailability assessment. Int J Nanomed 6:2429

Thakkar A, Chenreddy S, Wang J, Prabhu S (2015) Ferulic acid combined with aspirin demonstrates chemopreventive potential towards pancreatic cancer when delivered using chitosan-coated solid-lipid nanoparticles. Cell Biosci 5(1):1–14

Thangapazham RL, Puri A, Tele S, Blumenthal R, Maheshwari RK (2008) Evaluation of a nanotechnology-based carrier for delivery of curcumin in prostate cancer cells. Int J Oncol 32 (5):1119–1123

Van Duynhoven J, Vaughan EE, Jacobs DM, Kemperman RA, Van Velzen EJ, Gross G, Roger LC, Possemiers S, Smilde AK, Doré J, Westerhuis JA (2011) Metabolic fate of polyphenols in the human superorganism. Proc Natl Acad Sci 108(Supplement 1):4531–4538

Vijayakumar MR, Vajanthri KY, Balavigneswaran CK, Mahto SK, Mishra N, Muthu MS, Singh S (2016) Pharmacokinetics, biodistribution, in vitro cytotoxicity and biocompatibility of vitamin E TPGS coated trans resveratrol liposomes. Colloids Surf B: Biointerfaces 145:479–491

Vyas SP, Rai S, Paliwal R, Gupta PN, Khatri K, Goyal AK, Vaidya B (2008) Solid lipid nanoparticles (SLNs) as a rising tool in drug delivery science: one step up in nanotechnology. Curr Nanosci 4(1):30–44

Wu G, Li J, Yue J, Zhang S, Yunusi K (2018) Liposome encapsulated luteolin showed enhanced antitumor efficacy to colorectal carcinoma. Mol Med Rep 17(2):2456–2464

Yadav N, Parveen S, Banerjee M (2020) Potential of nano-phytochemicals in cervical cancer therapy. Clin Chim Acta 505:60–72

Yallapu MM, Khan S, Maher DM, Ebeling MC, Sundram V, Chauhan N, Ganju A, Balakrishna S, Gupta BK, Zafar N, Jaggi M (2014) Anti-cancer activity of curcumin loaded nanoparticles in prostate cancer. Biomaterials 35(30):8635–8648

Yang LJ, Ma SX, Zhou SY, Chen W, Yuan MW, Yin YQ, Yang XD (2013) Preparation and characterization of inclusion complexes of naringenin with β-cyclodextrin or its derivative. Carbohydr Polym 98(1):861–869

Zhang Y, Yang Y, Tang K, Hu X, Zou G (2008) Physicochemical characterization and antioxidant activity of quercetin-loaded chitosan nanoparticles. J Appl Polym Sci 107(2):891–897

5 Targeted Nanomedicine in Chemoprevention

Abstract

This chapter includes basics of targeted drug delivery and their advantages and disadvantages. The different strategies like drug conjugates and ligand anchored nanocarriers for targeted delivery of phytochemicals are also discussed. The next section demonstrates stimuli-responsive targeting such as pH-sensitive delivery and temperature responsive targeting of chemopreventive agents.

Keywords

Targeted delivery · Nanomedicine · Phytochemicals · Herbal drug · Drug targeting · Cancer cells

5.1 Introduction

Drug targeting using nanoparticles offers unique advantages such as controlled release at the diseased site, enhanced bioavailability in the tumor region, and less systemic side effects in the case of cytotoxic agents. Such precise control of site-selective drug delivery can be achieved through surface modification of nanoparticles using ligands, which through their interactions with respective receptors provide recognition and enhanced uptake of the cargos. Chemoprevention using such targeted nanomedicine explores the possibility of delivering phytochemicals to their selective target organ and target site within the intracellular environment. The literature of chemopreventive agents indicates organ specificity and hence superior efficacy of compounds like retinoids for breast, prostate, or urinary carcinomas than colon, similarly piroxicam and folic acid towards colon cancer than mammary cancers. Designing a targeted nanoformulation carrying such agents known for their natural specificity towards a specific organ through well-known over-expressed receptors in carcinoma of that organ may be a useful

R. Paliwal, S. R. Paliwal, *Advances in Nanochemoprevention*,
https://doi.org/10.1007/978-981-15-9692-6_5

approach for nanochemoprevention. Several ligands modified carriers have been used for achieving the advantages of targeted nanochemoprevention. The most commonly employed ligands for this purpose include folic acid, monoclonal antibodies, RGD peptide, hyaluronic acid, Apolipoprotein E3, estrogen, Octreotide peptide, HER-2, Fucose, and Breast tumor-homing cell-penetrating peptide, etc. (Paliwal et al. 2016; Rai et al. 2007).

5.2 Targeted Drug Delivery

Targeted drug delivery is necessary for cancer therapy as it involves the use of cytotoxic drugs that directly kill cells rather than its pharmacological effect on the cells through a biological response. Targeted delivery of drugs can be achieved by a wide variety of drug delivery systems. Targeting can be passive or active type, where passive targeting is the accumulation of drugs due to physicochemical or pharmacological factors and active targeting due to a specific interaction between ligand and receptor portals of the cell. The differential expression of tumor antigen and other receptor portals on cancer cells provide excellent opportunities for active targeting of anticancer drugs.

5.2.1 Passive Targeting

Drugs can be passively targeted to cancer cells through the leaky vasculature and accumulate there due to poor lymphatic drainage (enhanced permeation and retention effect) with the help of nanocarriers having high circulation time and sufficient hydrodynamic size to avoid renal filtration (Lee et al. 2006; Padilla De Jesús et al. 2002). The approximate size of fenestrations observed in leaky endothelium cells of tumor ranges between 10 nm and 100 nm, which also depends on the type of tumor, its nature, and the stage of malignancy (Hobbs et al. 1998, Hashizume et al. 2000). Thus nanocarriers of size ranging between 10 and 100 nm can be exploited for passive targeting of chemotherapeutic drugs based on the EPR effect (Gupta and Gupta 2005). Apart from size, the rapid clearance of nanocarriers *via* the reticuloendothelial system (RES) is also a matter of concern for passive targeting. To avoid RES uptake and prolong the circulation time, the surface of a delivery system is modified with hydrophilic surfactants/polymers or incorporating hydrophilic properties via hydrophilic biodegradable copolymers (e.g., polyethylene glycol (PEG), block copolymers of poly (ethylene oxide) (PEO), poloxamer, poloxamine, and Tween 80). The incorporation of such formulation changes decreases RES uptake and markedly prevents phagocytosis (Bazile et al. 1995) and reduces the RES uptake of nanocarriers facilitating effective passive tumor targeting (Gabizon and Papahadjopoulos 1988).

5.2.2 Active Targeting

In addition to passive targeting, active targeting via ligand for a specific surface receptor on the cancer cell provides numerous options for the development of effective anticancer therapeutics. Active targeting of the nanocarriers by attaching some target specific ligand and intracellular delivery strategy seems to improve the drug concentration at the target site and reduction of adverse effects on healthy cells.

Several ligands have been exploited for selective drug delivery, including antibodies, polypeptides, oligosaccharides, viral proteins, fusogenic peptides, and endogenous hormones, etc. (Paliwal et al. 2010). Although antibody appended drug delivery is regarded as a promising tool, some researcher has demonstrated that such strategy just increases internalization in animal models rather than effectively increasing localization of drugs in tumors (Bazak et al. 2015). Thus, to enhance cellular uptake of chemotherapeutic agent active targeting approach with molecular ligands is required. Internalization of drug nanocarriers by receptor-mediated endocytosis provided by specific targeted ligand molecules has shown to improve therapeutic efficacy (Steichen et al. 2013). Moreover, active targeting strategies are more significant for strategies such as gene delivery, gene silencing, etc., where internalization is the prerequisite for effective therapy (Ogris and Wagner 2002).

5.2.3 Drug Conjugates

The role of drug conjugates has been anticipated long back in drug targeting in order to achieve maximum therapeutic efficacy with minimum side effects (Ringsdorf 1975). Although drug conjugates have been successful in achieving objectives of targeted drug delivery; however, they suffer with complications of drug conjugation chemistry which may compromise the therapeutic efficacy of the drug molecules (Haag and Kratz 2006; Kim et al. 2009). For example, it is well known that hormones like Gonadotropin-releasing hormone (GnRH) and luteinizing hormone-releasing hormone (LHRH) are required in high concentrations for the development of pancreatic cancers cells; therefore, receptors for these hormone are over-expressed in these cells for tumor development, progression, and assistive vascularization. Limonta P and his group reported GnRHR targeted curcumin for targeted delivery to pancreatic cancer cells. They developed curcumin-DLys6-LHRH conjugate and studied their ex vivo and in vivo potential. The results demonstrated that the conjugate was equally useful to the drug alone but more soluble in water than free curcumin; hence can be used for intravenous administration and with lesser dose (Limonta et al. 2012). The conjugation of chemopreventive agents with targeted antitumor drugs improves their antitumor potential by increasing selectivity and decreasing non-specific distribution. Shi et al. (2012) developed flutamide and bicalutamide conjugated curcumin for the targeted delivery of these antiandrogens to prostate cancer cells with higher efficacy with lesser side effects. The curcumin conjugates have been more effective than the drug alone (Shi et al. 2012). The antibody-conjugated chemopreventive agents have also been a well-accepted and

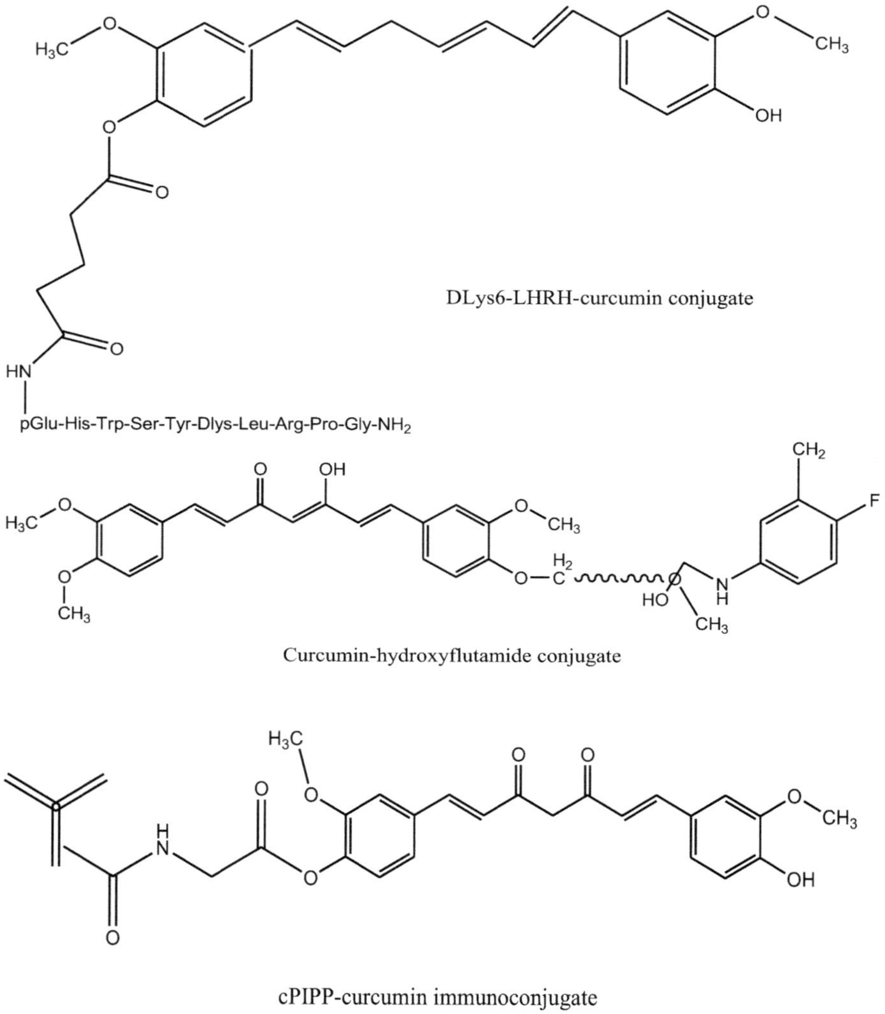

Fig. 5.1 Chemical structure of curcumin hybrid compounds with anticancer properties. LHRH, luteinizing hormone-releasing hormone (Adopted from Teiten et al. 2014)

explored strategy for targeted delivery to cancer cells. Vyas and his group reported curcumin immunoconjugate for the targeted delivery to hCGβ expressing sites by myeloid leukemia cells (Vyas et al. 2009). Though the drug conjugates increase the therapeutic potential of cytotoxic agents, sometimes they did not improve the therapeutic properties of a lined moiety like curcumin-paclitaxel conjugate (Nakagawa-Goto et al. 2007) (Fig. 5.1).

5.2.4 Ligand Anchored Nanocarriers Mediated Targeting

Generally, nanocarriers are used to improve the bioavailability of chemopreventive agents by encapsulating them within the nanocarriers, and they also increase the solubility of poorly water-soluble drugs. They maintain the therapeutic level for a prolonged period and can also be used for targeted drug delivery by utilizing targeting ligand. Several ligand anchored nanocarriers have been reported for the targeted delivery of chemopreventive agents with improved bioavailability (Table 5.1).

Prostate-specific membrane antigen is a type-II transmembrane protein that is present in prostate tissue. It is widely explored for diagnosis and targeted drug delivery to prostate cancer cells. Prostate membrane specific antigen (PSMA) targeted formulations have been widely explored in the literature. A PSMA specific antibody appended liposomal formulation carrying curcumin showed 10-times lower dose in achieving about 80% inhibition of cellular proliferation (Thangapazham et al. 2008). In a similar study, Sanna et al. (2011) developed EGCG loaded PEGylated PLGA nanoparticles anchored with a peptide possessing specificity towards PSMA over-expressed LNCaP cells. A higher cellular binding, along with increased antiproliferative activity, was reported by authors (Sanna et al. 2011). For targeting tumor vasculature, RGD peptide serves as a ligand in cancer therapy. RGD peptide conjugated PEG–PLA micelles having curcumin showed a stronger inhibiting effect on the growth of tumors compared with non-RGD modified micelles (Zhao et al. 2015).

Over-expressed CD44 receptors can be easily targeted using hyaluronic acid as a ligand. Recently, Kesharwani et al. (2015) synthesized hyaluronic acid conjugate of the copolymer (styrene–maleic acid) (HA-SMA) micelles carrying a curcumin derivative, i.e., 3, 4-difluorobenzylidene curcumin for CD44 targeting. Such micelles have shown significant implications ex vivo in treating pancreatic cancers, including the more aggressive pancreatic CSLCs (Kesharwani et al. 2015). Cholesteryl-hyaluronic acid nanogel of curcumin has been found 13-fold higher effective in tumor suppression than free drug (Wei et al. 2014). Lin and co-workers developed folic acid-containing polyethylene glycol (PEG)–distearoylphosphatidylethanolamine nanoparticles for selective delivery of curcumin to cancer cells over-expressing FA receptors (Lin et al. 2015). In a similar study, genistein was incorporated into cellulose nanocrystals anchored with folic acid to achieve folate receptor-mediated cellular binding/uptake of the conjugate on humans (DBTRG-05MG, H4) and rat (C6) brain tumor cells (Dong et al. 2014). In a study, folic acid anchored PLGA–PEG nanoparticles carrying a camptothecin analog SN-38 showed significantly higher cytotoxicity (Ebrahimnejad et al. 2010).

The curcumin-difluorinated is a synthetic analog of curcumin that possesses high anticancer activity and increased half-life in comparison to curcumin. It shows high metabolic stability, growth-inhibitory properties, and multidrug resistance reversal properties by inhibiting ATP binding cassette subfamily (Kanwar et al. 2011). Alsaab and co-workers reported folic acid targeted micelles were also developed for the delivery of hydrophobic curcumin-difluorinated for retinoblastoma cells. The

Table 5.1 Enlisted targeted nanomedicine developed for site-specific delivery of chemopreventive agents

Nanocarrier	Bioactive	Targeting ligand	Remarks	Reference
Liposomes	Curcumin	Prostate membrane specific antigen-specific antibodies	The PSMA targeted liposomal curcumin (5–10 μM) resulted in 70–80% inhibition of cellular proliferation in 24 h. However for similar inhibition tenfold higher doses of free curcumin required	Thangapazham et al. (2008)
PLGA–PEG–folate nanoparticles	Camptothecin analog SN-38	Folic acid	Significantly higher cytotoxicity was observed	Ebrahimnejad et al. (2010)
Fe_3O_4 nanoparticles coated by cross-linked carboxymethylated chitosan nanoparticles	Genistein	Magnetic hyperthermia	Multifunctional chemotherapeutic application combined with drug release and magnetic hyperthermia	Si et al. (2010)
Poly(lactide-co-glycolide)-PEG, PLGA–PEG] nanoparticles	EGCG	Pseudomimetic dipeptide *N*-[*N*-[(S)-1,3-dicarboxypropyl] Carbamoyl]-(S)-lysine (DCL)	The targeted nanoparticles showed specificity towards prostate membrane specific antigen (PSMA), and significantly enhanced the binding, and exhibited an increased antiproliferative activity against PSMA-expressing LNCaP cells with respect to the non-functionalized ones, without affecting normal cell viability	Sanna et al. (2011)
Dendritic Boltorn® H40 core, a hydrophobic poly(l-lactide) (PLA) inner shell, and a hydrophilic poly (ethylene glycol) (PEG) outer shell micelles	Resveratrol	Octreotide peptide	Micelle may offer a promising approach for targeted neuroendocrine tumor (NET) therapy that is often resistant to standard therapies	Xu et al. (2013)
Immunoliposome	Curcumin and resveratrol	HER-2	Immunoliposome formulation containing both resveratrol and curcumin suggests a multi-targeted mechanism of action	Catania et al. (2013)

PEG–PLA micelles	Curcumin	αvβ3 integrin-targeted peptide (RGD)	The targeted formulation showed a more substantial inhibiting effect on the growth of tumors compared with non-RGD modified micelles	Zhao et al. (2015)
Cholesteryl-hyaluronic acid nanogel	Curcumin	Hyaluronic acid	Resulted in up to 13-fold tumor suppression	Wei et al. (2014)
Cellulose nanocrystals	Genistein	Folic acid	Folate receptor-mediated cellular binding/uptake of the conjugate was demonstrated on human (DBTRG-05MG, H4) and rat (C6) brain tumor cells	Dong et al. (2014)
Transferrin-targeted PEG–PE micelles	Curcumin and paclitaxel	Transferrin	The formulation was suitable for the treatment of resistant ovarian cancer	Abouzeid et al. (2014)
Phospholipids with the low-density lipoprotein receptor (LDLr)-binding domain of apolipoprotein E3	Resveratrol	Apolipoprotein E3	rHDL provides an ideal hydrophobic milieu to sequester resveratrol and that rHDL containing apoE3 serves as an effective "nanovehicle" to transport and deliver resveratrol to targeted intracellular sites	Kim et al. (2015)
The hyaluronic acid conjugate of copoly(styrene–maleic acid) (HA-SMA) micelles	3,4-difluorobenzylidene curcumin	Hyaluronic acid	CD44 targeted nanomicelles may have significant implications in treating pancreatic cancers including the more aggressive pancreatic CSLCs	Kesharwani et al. (2015)
Folic acid-containing polyethylene glycol (PEG)–distearoylphosphatidylethanolamine nanoparticles	Curcumin	Folic acid	Selectively deliver the drug to cancer cells over-expressing FA receptors	Lin et al. (2015)

(continued)

Table 5.1 (continued)

Nanocarrier	Bioactive	Targeting ligand	Remarks	Reference
Fucose-conjugated chitosan and polyethylene glycol-conjugated chitosan complex with gelatin	Epigallocatechin-3-gallate	Fucose	Nanoparticles effectively reduced drug release within gastric acids and that a controlled epigallocatechin-3-gallate release inhibited gastric cancer cell growth, induced cell apoptosis, and reduced vascular endothelial growth factor protein expression	Lin et al. (2015)
Colloidal mesoporous silica nanoparticles	Epigallocatechin-3-gallate	Breast tumor-homing cell-penetrating peptide (PEGA–pVECpeptide)	Peptide significantly promotes the efficacy of EGCG on breast tumors by targeted accumulation and release	Ding et al. (2015)
pH-sensitive β-cyclodextrin nanoparticles	Resveratrol	pH-sensitive approach	Effective tumor growth inhibition was observed.	Lv et al. (2016)
Folic acid targeted micelles	Curcumin-difluorinated	Folic acid	Significant cytotoxic potential on Y-79 and WERI-RB cell lines was reported	Alsaab et al. (2017)
Folate targeted PLGA nanoparticles	Boron–curcumin and gadolinium	Folic acid	Combination therapy was found superior on ovarian cancer (IGROV-1) cells	Alberti et al. (2017)
Folic acid targeted *β*-cyclodextrin linked magnetic nanoparticles	Curcumin	Folic acid	Enhanced uptake on HepG2 cells	Song et al. (2018)

micellar formulation revealed higher drug loading and significant cytotoxic potential on Y-79 and WERI-RB cell lines. Simultaneously, the formulation did not reveal any cytotoxicity on human retinal pigment epithelial cell (Alsaab et al. 2017). An apolipoprotein E3 mediated delivery of resveratrol entrapped into hydrophobic core of phospholipids of high-density lipoproteins shown its effective transport and delivery to targeted intracellular sites (Kim et al. 2015).

Micellar system having a hydrophobic poly(l-lactide) (PLA) inner shell, and a hydrophilic poly(ethylene glycol) (PEG) outer shell modified with octreotide peptide for targeted delivery of resveratrol to neuroendocrine tumor has been developed (Xu et al. 2013). It is claimed that the system will be effective even in resistant tumors too. HER-2 anchored immunoliposomal formulation containing both resveratrol and curcumin showed a multi-targeted mechanism of action and better efficacy in chemoprevention (Catania et al. 2013).

Fucose as a ligand has been conjugated with chitosan complexed with another biopolymer gelatin. Targeted nanocomplex loaded with EGCG showed targeted and controlled delivery to cancer cells (Lin et al. 2015). The system was capable of reducing drug release significantly in gastric acids environment. When tested against gastric cancer cell, nanocomplex driven controlled EGCG release contributed to growth inhibition, induced cell apoptosis, and reduced vascular EGFR protein expression. Breast tumor-homing cell-penetrating peptide anchored colloidal mesoporous silica nanoparticles for EGCG delivery greatly promoted the efficacy of EGCG on breast tumors by targeted accumulation and release (Ding et al. 2015).

Hepatocyte targeted chemopreventive agents can be developed by conjugating galactose to a delivery system which has specificity for asialoglycoprotein receptors on hepatoma cells. The ligands with d-galactose and *N*-acetylgalactosamine terminals have been widely exploited for this purpose (Wang et al. 2006; Wu et al. 2002). Zhou and co-workers examined targeting potential of curcumin loaded galactose anchored chitosan polycaprolactone nanoparticles for the hepatic cells (Zhou et al. 2013). They reported the enhanced accumulation of targeted formulation as compared to free curcumin in hepatocellular carcinoma (HepG2) cells, with approximately sixfold higher potential to induce apoptosis and necrosis.

5.3 Stimuli-Responsive Targeted Drug Delivery

For an effective treatment strategy, the nanocarriers must release the drug at the site of action at the desired rate. The drug release from nanocarriers can be controlled by utilizing specific features of carcinogenic cells such as pH, temperature, improved glycolysis, or presence of proteases. Considering these points, several stimuli-responsive drug delivery systems have been developed. The low pH of the tumor environment is important and has been widely exploited for pH-responsive targeted drug delivery to the tumor tissues (Paliwal et al. 2015). The pH-responsive nanocarriers have been extensively used for selective delivery to the cancer site. The tumor extracellular environment is acidic in comparison to surrounding healthy tissues due to higher metabolism and scarcity of oxygen (Tan et al. 2015). These pH-sensitive nanocarriers protect the encapsulated drug until they reached to tumor site via EPR effect or navigated through targeting ligand, where loaded drug is released due to destabilization of nanocarrier system (Paliwal et al. 2012). Lv et al. (2016) reported pH-sensitive resveratrol loaded β-cyclodextrin nanoparticles entrapped in microbubbles for targeted drug delivery. The developed system was efficiently suppressed the tumor growth (Lv et al. 2016).

Thermoresponsive targeting is the site-specific delivery of nanocarriers on the basis of hyperthermia at the cancer sites. Thermoresponsive materials are used for the formulation of such nanocarriers, which protect the encapsulated content in systemic circulation, and drug is released at the cancer site. Reijnold and co-workers reported thermoresponsive chitosan-based nanoparticles for site-specific delivery to solid tumors. As discussed earlier, targeted drug delivery can be achieved by anchoring various ligands that are specific for the cell surface receptors. Magnetic targeting is also widely used for this purpose. Lübbe et al. (1996) reported magnetic nanoparticles for drug delivery and are at present approved for pharmaceutical application. In such case, drug has been conjugated with the magnetic carrier, which navigates the drug under the influence of magnetic field to the target site. Dai Tran et al. (2010) reported magnetic curcumin chitosan magnetofluorescent conjugate as targeted nanotheranostic. In another study, magnetic silk fibroin core nanoparticles were used for targeted delivery of curcumin to breast cancer cells. The breast cancer cells MDA-MB-231 showed higher accumulation of these nanoparticles with enhanced cytotoxic potential than curcumin alone (Song et al. 2017). Shengmei and co-workers prepared folic acid targeted *β*-cyclodextrin linked magnetic nanoparticles for controlled and targeted delivery of curcumin (Song et al. 2018). The developed formulation was non-toxic, biocompatible, and showed enhanced uptake on HepG2 cells and suggested the promising applicability of such a system for site-specific delivery of the water-insoluble chemopreventive agent for prevention and cancer therapy. Magnetic nanoparticles coated with cross-linked carboxymethylated chitosan for controlled delivery and magnetic hyperthermia of genistein have been developed for multifunctional chemotherapeutic application (Si et al. 2010). Multifunctional nanocarriers can be used as a single agent for controlled and targeted drug delivery along with diagnostic applications as theranostic agents (Jabr-Milane et al. 2008). Alberti et al. (2017) reported folate targeted PLGA nanoparticles loaded with boron–curcumin and gadolinium for dual application of imaging as well as anticancer therapy.

5.4 Conclusion

The phytochemicals are beneficial but are associated with some drawbacks such as non-specificity and low bioavailability. The application of a novel drug delivery system improves bioavailability to some extent; however, to further enhance the potential of these agents, several targeting strategies have been exploited. The over-expressed receptors have been successfully utilized for targeted drug delivery. The above discussed reports indicate the excellent utility of such targeted nanomedicine or "nano-death-packets" designed and developed exclusively for chemoprevention using natural compounds, nanotechnology, and bio-molecular recognition through ligand–receptor interactions.

References

Abouzeid AH, Patel NR, Sarisozen C, Torchilin VP (2014) Transferrin-targeted polymeric micelles co-loaded with curcumin and paclitaxel: efficient killing of paclitaxel-resistant cancer cells. Pharm Res 31(8):1938–1945

Alberti D, Protti N, Franck M, Stefania R, Bortolussi S, Altieri S, Deagostino A, Aime S, Geninatti Crich S (2017) Theranostic nanoparticles loaded with imaging probes and rubrocurcumin for combined cancer therapy by folate receptor targeting. ChemMedChem 12(7):502–509

Alsaab H, Alzhrani RM, Kesharwani P, Sau S, Boddu SH, Iyer AK (2017) Folate decorated nanomicelles loaded with a potent curcumin analogue for targeting retinoblastoma. Pharmaceutics 9(2):15

Bazak R, Houri M, El Achy S, Kamel S, Refaat T (2015) Cancer active targeting by nanoparticles: a comprehensive review of literature. J Cancer Res Clin Oncol 141(5):769–784

Bazile D, Prud'homme C, Bassoullet MT, Marlard M, Spenlehauer G, Veillard M (1995) Stealth Me. PEG-PLA nanoparticles avoid uptake by the mononuclear phagocytes system. J Pharm Sci 84(4):493–498

Catania A, Barrajón-Catalán E, Nicolosi S, Cicirata F, Micol V (2013) Immunoliposome encapsulation increases cytotoxic activity and selectivity of curcumin and resveratrol against HER2 overexpressing human breast cancer cells. Breast Cancer Res Treat 141(1):55–65

Dai Tran L, Hoang NMT, Mai TT, Tran HV, Nguyen NT, Tran TD, Do MH, Nguyen QT, Pham DG, Ha TP, Van Le H (2010) Nanosized magnetofluorescent Fe3O4–curcumin conjugate for multimodal monitoring and drug targeting. Colloids Surf A Physicochem Eng Asp 371 (1-3):104–112

Ding J, Yao J, Xue J, Li R, Bao B, Jiang L, Zhu JJ, He Z (2015) Tumor-homing cell-penetrating peptide linked to colloidal mesoporous silica encapsulated (−)-epigallocatechin-3-gallate as drug delivery system for breast cancer therapy in vivo. ACS Appl Mater Interfaces 7 (32):18145–18155

Dong S, Cho HJ, Lee YW, Roman M (2014) Synthesis and cellular uptake of folic acid-conjugated cellulose nanocrystals for cancer targeting. Biomacromolecules 15(5):1560–1567

Ebrahimnejad P, Dinarvand R, Sajadi A, Jaafari MR, Nomani AR, Azizi E, Rad-Malekshahi M, Atyabi F (2010) Preparation and in vitro evaluation of actively targetable nanoparticles for SN-38 delivery against HT-29 cell lines. Nanomed Nanotechnol Biol Med 6(3):478–485

Gabizon A, Papahadjopoulos D (1988) Liposome formulations with prolonged circulation time in blood and enhanced uptake by tumors. Proc Natl Acad Sci 85(18):6949–6953

Gupta AK, Gupta M (2005) Synthesis and surface engineering of iron oxide nanoparticles for biomedical applications. Biomaterials 26(18):3995–4021

Haag R, Kratz F (2006) Polymer therapeutics: concepts and applications. Angew Chem Int Ed 45(8):1198–1215

Hashizume H, Baluk P, Morikawa S, McLean JW, Thurston G, Roberge S, Jain RK, McDonald DM (2000) Openings between defective endothelial cells explain tumor vessel leakiness. Am J Pathol 156(4):1363–1380

Hobbs SK, Monsky WL, Yuan F, Roberts WG, Griffith L, Torchilin VP, Jain RK (1998) Regulation of transport pathways in tumor vessels: role of tumor type and microenvironment. Proc Natl Acad Sci 95(8):4607–4612

Jabr-Milane LS, van Vlerken LE, Yadav S, Amiji MM (2008) Multi-functional nanocarriers to overcome tumor drug resistance. Cancer Treat Rev 34(7):592–602

Kanwar SS, Yu Y, Nautiyal J, Patel BB, Padhye S, Sarkar FH, Majumdar AP (2011) Difluorinated-curcumin (CDF): a novel curcumin analog is a potent inhibitor of colon cancer stem-like cells. Pharm Res 28(4):827–838

Kesharwani P, Banerjee S, Padhye S, Sarkar FH, Iyer AK (2015) Hyaluronic acid engineered nanomicelles loaded with 3, 4-difluorobenzylidene curcumin for targeted killing of CD44+ stem-like pancreatic cancer cells. Biomacromolecules 16(9):3042–3053

Kim S, Kim JH, Jeon O, Kwon IC, Park K (2009) Engineered polymers for advanced drug delivery. Eur J Pharm Biopharm 71(3):420–430

Kim SH, Adhikari BB, Cruz S, Schramm MP, Vinson JA, Narayanaswami V (2015) Targeted intracellular delivery of resveratrol to glioblastoma cells using apolipoprotein E-containing reconstituted HDL as a nanovehicle. PLoS One 10(8):e0135130

Lee CC, Gillies ER, Fox ME, Guillaudeu SJ, Fréchet JM, Dy EE, Szoka FC (2006) A single dose of doxorubicin-functionalized bow-tie dendrimer cures mice bearing C-26 colon carcinomas. Proc Natl Acad Sci 103(45):16649–16654

Limonta P, Marelli MM, Mai S, Motta M, Martini L, Moretti RM (2012) GnRH receptors in cancer: from cell biology to novel targeted therapeutic strategies. Endocr Rev 33(5):784–811

Lin YH, Chen ZR, Lai CH, Hsieh CH, Feng CL (2015) Active targeted nanoparticles for oral administration of gastric cancer therapy. Biomacromolecules 16(9):3021–3032

Lübbe AS, Bergemann C, Huhnt W, Fricke T, Riess H, Brock JW, Huhn D (1996) Preclinical experiences with magnetic drug targeting: tolerance and efficacy. Cancer Res 56 (20):4694–4701

Lv Y, Hao L, Hu W, Ran Y, Bai Y, Zhang L (2016) Novel multifunctional pH-sensitive nanoparticles loaded into microbubbles as drug delivery vehicles for enhanced tumor targeting. Sci Rep 6:29321

Nakagawa-Goto K, Yamada K, Nakamura S, Chen TH, Chiang PC, Bastow KF, Wang SC, Spohn B, Hung MC, Lee FY, Lee FC (2007) Antitumor agents. 258. Syntheses and evaluation of dietary antioxidant—taxoid conjugates as novel cytotoxic agents. Bioorg Med Chem Lett 17 (18):5204–5209

Ogris M, Wagner E (2002) Targeting tumors with non-viral gene delivery systems. Drug Discov Today 7(8):479–485

Padilla De Jesús OL, Ihre HR, Gagne L, Fréchet JM, Szoka FC (2002) Polyester dendritic systems for drug delivery applications: in vitro and in vivo evaluation. Bioconjug Chem 13(3):453–461

Paliwal SR, Paliwal R, Mishra N, Mehta A, Vyas SP (2010) A novel cancer targeting approach based on estrone anchored stealth liposome for site-specific breast cancer therapy. Curr Cancer Drug Targets 10(3):343–353

Paliwal SR, Paliwal R, Pal HC, Saxena AK, Sharma PR, Gupta PN, Agrawal GP, Vyas SP (2012) Estrogen-anchored pH-sensitive liposomes as nanomodule designed for site-specific delivery of doxorubicin in breast cancer therapy. Mol Pharm 9(1):176–186

Paliwal SR, Paliwal R, Vyas SP (2015) A review of mechanistic insight and application of pH-sensitive liposomes in drug delivery. Drug Delivery 22(3):231–242

Paliwal SR, Paliwal R, Agrawal GP, Vyas SP (2016) Hyaluronic acid modified pH-sensitive liposomes for targeted intracellular delivery of doxorubicin. J Liposome Res 26(4):276–287

Rai S, Paliwal R, Vaidya B, Gupta PN, Mahor S, Khatri K, Goyal AK, Rawat A, Vyas SP (2007) Estrogen (s) and analogs as a non-immunogenic endogenous ligand in targeted drug/DNA delivery. Curr Med Chem 14(19):2095–2109

Ringsdorf H (1975) Structure and properties of pharmacologically active polymers. J Polym Sci Polym Symp 51(1):135–153

Sanna V, Pintus G, Roggio AM, Punzoni S, Posadino AM, Arca A, Marceddu S, Bandiera P, Uzzau S, Sechi M (2011) Targeted biocompatible nanoparticles for the delivery of (−)-epigallocatechin 3-gallate to prostate cancer cells. J Med Chem 54(5):1321–1332

Shi Q, Wada K, Ohkoshi E, Lin L, Huang R, Morris-Natschke SL, Goto M, Lee KH (2012) Antitumor agents 290. Design, synthesis, and biological evaluation of new LNCaP and PC-3 cytotoxic curcumin analogs conjugated with anti-androgens. Bioorg Med Chem 20 (13):4020–4031

Si HY, Li DP, Wang TM, Zhang HL, Ren FY, Xu ZG, Zhao YY (2010) Improving the anti-tumor effect of gbenistein with a biocompatible superparamagnetic drug delivery system. J Nanosci Nanotechnol 10(4):2325–2331

Song W, Muthana M, Mukherjee J, Falconer RJ, Biggs CA, Zhao X (2017) Magnetic-silk core–shell nanoparticles as potential carriers for targeted delivery of curcumin into human breast cancer cells. ACS Biomater Sci Eng 3(6):1027–1038

Song S, Li M, Gong X, Han H, Zhou Y, Wang L, Shuang S, Dong C (2018) Controlled release of curcumin via folic acid conjugated magnetic drug delivery system. Chem Res Chin Univ 34(2):203–211

Steichen SD, Caldorera-Moore M, Peppas NA (2013) A review of current nanoparticle and targeting moieties for the delivery of cancer therapeutics. Eur J Pharm Sci 48(3):416–427

Tan S, Wu T, Zhang D, Zhang Z (2015) Cell or cell membrane-based drug delivery systems. Theranostics 5(8):863

Teiten MH, Dicato M, Diederich M (2014) Hybrid curcumin compounds: a new strategy for cancer treatment. Molecules 19(12):20839–20863

Thangapazham RL, Puri A, Tele S, Blumenthal R, Maheshwari RK (2008) Evaluation of a nanotechnology-based carrier for delivery of curcumin in prostate cancer cells. Int J Oncol 32 (5):1119–1123

Vyas HK, Pal R, Vishwakarma R, Lohiya NK, Talwar GP (2009) Selective killing of leukemia and lymphoma cells ectopically expressing HCGβ by a conjugate of curcumin with an antibody against HCGβ subunit. Oncology 76(2):101–111

Wang SN, Deng YH, Xu H, Wu HB, Qiu YK, Chen DW (2006) Synthesis of a novel galactosylated lipid and its application to the hepatocyte-selective targeting of liposomal doxorubicin. Eur J Pharm Biopharm 62(1):32–38

Wei X, Senanayake TH, Bohling A, Vinogradov SV (2014) Targeted nanogel conjugate for improved stability and cellular permeability of curcumin: synthesis, pharmacokinetics, and tumor growth inhibition. Mol Pharm 11(9):3112–3122

Wu J, Nantz MH, Zern MA (2002) Targeting hepatocytes for drug and gene delivery: emerging novel approaches and applications. Front Biosci 7:d717–d725

Xu W, Burke JF, Pilla S, Chen H, Jaskula-Sztul R, Gong S (2013) Octreotide-functionalized and resveratrol-loaded unimolecular micelles for targeted neuroendocrine cancer therapy. Nanoscale 5(20):9924–9933

Zhao L, Yang C, Dou J, Xi Y, Lou H, Zhai G (2015) Development of RGD-functionalized PEG-PLA micelles for delivery of curcumin. J Biomed Nanotechnol 11(3):436–446

Zhou N, Zan X, Wang Z, Wu H, Yin D, Liao C, Wan Y (2013) Galactosylated chitosan–polycaprolactone nanoparticles for hepatocyte-targeted delivery of curcumin. Carbohydr Polym 94(1):420–429

Miscellaneous Approaches of Chemoprevention

6

Abstract

This chapter deals with the miscellaneous approaches of chemoprevention including synthetic chemicals, vitamins, NSAIDs, antioxidants, etc. Role of silver nanoparticles and selenium nanoparticles is also described in chemoprevention. A section on combination therapies using both phytopharmaceuticals and synthetic anticancer drugs is also added. The reports on simultaneous use of diagnostic agent with chemopreventive agents are further described in the chapter.

Keywords

Combination cancer therapy · Silver nanoparticles · Selenium nanoparticles · Herbal anticancer medicine

6.1 Introduction

Chemoprevention includes use of either a medication or a vitamin or a dietary supplement in order to avoid cancer. Chemopreventive medicinal agents are broadly classified in to phytopharmaceuticals based compounds and non-phytopharmaceuticals compounds. Apart from phytochemicals, chemoprevention can also be achieved by several miscellaneous approaches that include use of silver nanoparticles, combination strategies, use of dietary compounds, NSAIDs, antioxidants, etc. Like natural molecules, synthetic drugs have also shown remarkable increase in their potency upon encapsulation in nanoparticles. In this chapter, we will discuss these approaches of chemoprevention.

R. Paliwal, S. R. Paliwal, *Advances in Nanochemoprevention*,
https://doi.org/10.1007/978-981-15-9692-6_6

6.2 Silver Nanoparticles As Novel Chemopreventive Agents

Silver nanoparticles (AgNPs) have been perceived as promising agents and are broadening their utilization in diagnosis and treatment. Silver possesses free radical scavenging, antimicrobial, and anti-inflammatory activity and has been used to prevent and treat number of diseases (Bhol et al. 2004; Jain et al. 2009). For these pharmacological activities, AgNPs are available in variety of pharmaceutical products such as ointments, bandages, wound dressings (Jones et al. 2004; Ip et al. 2006). The commercial availability of number of silver nanoparticles based formulations suggests its applicability and safety (Kokura et al. 2010; Munger et al. 2014). Along with these effects, AgNPs also possess cytotoxic potential against various cancers (Gengan et al. 2013; Jeyaraj et al. 2013a, b). Several mechanisms have been reported for AgNPs activity including apoptosis mediated (Hsin et al. 2008), anti-angiogenic activity (Gurunathan et al. 2009), DNA damage, and caspase-3 mediated cell death (Jeyaraj et al. 2013a, b). According to "World Cancer Report," skin cancer incidence is about 30% of all the diagnosed cancer (Lazovich et al. 2012). This scenario demands for an effective and safe formulation that reduces the occurrence of skin carcinogenesis. Arora and co-workers studied the potential of AgNPs for skin protection from UV rays leading to DNA damage (Arora et al. 2015). Authors claimed that AgNPs have capability in reducing apoptosis and inducing cell cycle arrest (G1/S phase) in human immortalized keratinocytes. It was also noted that AgNPs were involved in DNA damage repair.

6.3 Non-Steroidal Anti-inflammatory Drugs for Chemoprevention

A large number of studies reported that COX-2 enzymes are over-expressed in cancerous cells (Chan et al. 1999) and are therefore involved in number of process of carcinogenesis such as inhibition of apoptosis, promotion of cell proliferation, and angiogenesis (Marnett 1994; Tsujii and DuBois 1995). Non-steroidal anti-inflammatory drugs (NSAIDs), more specifically specific COX-2 inhibitors, prevent carcinogenesis by blocking COX-2 dependent processes. A number of pre-clinical and clinical studies demonstrated the potential role of NSAIDs in cancer chemoprevention. NSAIDs act by inhibiting enzyme cyclooxygenase, which is responsible for the synthesis of prostaglandin, a mediator in the process of inflammation (Subbaramaiah and Dannenberg 2003). NSAIDs are classified in two categories, COX-1 and COX-2 isoforms. The use of NSAIDs such as aspirin and selective cyclooxygenase-2 (COX-2) inhibitors has been shown effective in the reduction of risk of various cancers (Rao and Reddy 2004). However, they also act on other COX-2 independent molecular pathways that include PPAR receptors, NFκB and apoptosis mediated by BAX, etc. (Soh and Weinstein 2003). Figure 6.1 represents the summary of the COX-2 dependent and independent actions of NSAIDs.

The disturbance in programmed cell death (apoptosis) mechanism, which upsets the cell growth and death equilibrium, is a significant contributor in cancer

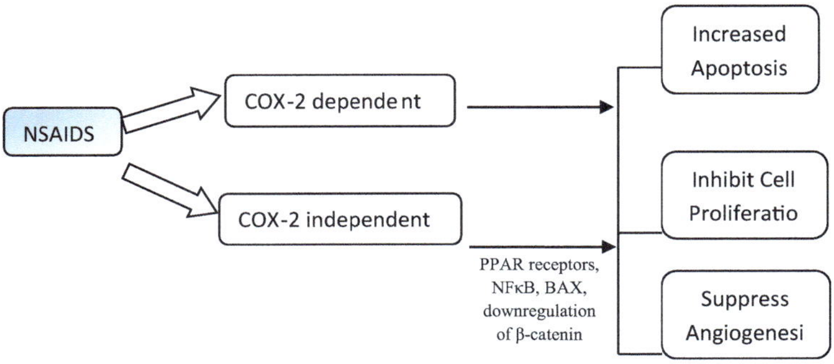

Fig. 6.1 COX-2 dependent and independent action of NSAIDs

development (Lowe et al. 2004). Apart from COX inhibition, NSAIDs exert pro-apoptotic properties on cancer cells (Seo et al. 2007). This prompted the investigation of remedial enactment of apoptosis in malignant growth cells as a possible chemoprevention strategy (Reed 2006). The COX-2 independent chemopreventive activities of NSAIDs include inhibition of NF-κB (Takada et al. 2004), cellular adhesion (Tozawa et al. 1995), downregulation of β-catenin (Koornstra et al. 2005), and downregulation of EGF receptor signaling (Pangburn et al. 2010). The COX-independent actions may possibly permit advancement of NSAIDs with chemotherapeutic adequacy while avoiding the untoward impacts due to COX inhibition. Thakkar et al. (2015a, b) reported ibuprofen loaded solid lipid nanoparticles as novel chemopreventive formulation. The developed SLNs were more efficacious with tenfold reduction in IC50 concentration and fourfold reduction in dose in comparison to free drug (Thakkar et al. 2015a, b).

6.4 Selective Antioxidants and Chemoprevention

6.4.1 Selenium

Selenium, a trace element, has excellent physiological and pharmacological features such as reduction of diseases like diabetes, cardiovascular, neurodegenerative disorder including cancer. It has transitional characters from metal to non-metal and organic to inorganic forms. Selenium can take many shapes such as nanospheres, nanotubes, nanorods, and nanowires. Selenium exerts inhibitory effect on variety of cancers through regulation of apoptosis (Zheng et al. 2012; Wang et al. 2014; Gao et al. 2014; Zhou et al. 2016; Liao et al. 2016). It is also reported that SeNPs are biodegradable in vivo and serve as nutrient for normal cells and exert cytotoxicity for cancerous cells (Zheng et al. 2012). Recently, it was reviewed that selenium nanoparticles (SeNPs) can act as a pro-oxidant or antioxidant agent (Menon and Shanmugam 2020). The surface functionalization of SeNPs further enhances its chemotherapeutic effect and may be presented as novel nanodrugs with less toxicity, high biocompatibility, and better reactivity. The possible mechanisms associated with SeNPs may be ROS production, apopototic pathway activation, and caspase cascade or cell cycle arrest.

6.4.2 α-Tocopherol

Vitamin E, a fat soluble vitamin, is an excellent antioxidant source with total eight variants including four tocopherols (α-, β-, γ-, and δ-tocopherols) and four tocotrienols (α-, β-, γ-, and δ-tocotrienols). Among all, alpha-tocopherol is a major variant and has antioxidant properties (Ekstrand-Hammarström et al. 2007; Constantinou et al. 2008; Smolarek and Suh 2011; Smolarek et al. 2012). The α-tocopherol or vitamin E is an effective agent which is required in promotional phase of carcinogenesis. The promotional phase is associated with cellular damage resulted by reactive oxygen species and α-tocopherol protects the cells by such damage (Summerfield and Tappel 1984). It also lowers the level of carcinogenic nitrosamines by inhibiting nitrosation reactions (Bartsch et al. 1983).

6.4.3 Vitamin A: The Retinoids and Carotenoids

The retinoids and carotenoids are group of compounds that also demonstrated chemopreventive and anti-carcinogenic potential. The dietary carotenoid, AY carotene is an important component for chemopreventive action (Peto et al. 1981). The retinoid regulates the growth and differentiation of normal and cancer cells. They also restrain the effect of various carcinogens and stimulate apoptosis (Trump 1994). These effects are regulated by two nuclear receptors families which are retinoid acid receptors and retinoid X receptors. The disturbance in the retinoid metabolism is reported in number of cancers (Kristal 2004) and restoration of expression of these receptors via administration of dietary vitamin A can be used as chemopreventive strategy (Lotan et al. 2000). The carotenoids like β-carotene also possess chemopreventive activity (Tanaka et al. 2012).

6.4.4 Other Antioxidants

Activated oxygen species are involved in carcinogenesis by two mechanisms; direct mechanism, which damages DNA base or indirect mechanism where lipid peroxidation is involved. Therefore, other antioxidants such as butylated hydroxytoluene (BHT), ascorbic acid, and other dietary antioxidants like β-carotene can be used to reduce the carcinogenesis.

6.5 Combination Therapy

Combination therapy offers advantages over single therapy in the sense that it maximizes the efficacy of the formulation and minimizes the potential side effects of single dose therapy. Nowadays, more research is being focused on such combination therapies based on the development of plant based agents to acquire clinical effectiveness. Recent studies in the area of chemoprevention have proved that combination therapy is more useful and successfully increased efficacy synergistically by acting on multiple signaling pathways with minimal toxic effects (Narayanan et al. 2004). This can be ascribed to the multi-factorial complex nature of carcinogenesis wherein malignancy is the result of different cell changes. Combined use of phytochemicals with either anticancer drugs or other chemopreventive agents or diagnostic agents has been used in literature and is discussed in coming sections. Table 6.1 summarizes some of the reports published using nanoparticles loaded with miscellaneous chemopreventive agents.

6.5.1 Phytochemicals and Antineoplastic Agents

Combination of phytochemicals with classical anticancer agents like doxorubicin can reduce the doses and associated side effects. Mahbub et al. (2015) reported that low dose of doxorubicin and flavonoids combination regimen have been synergistically acted on malignant cells. They effectively reduced ATP level, induced apoptosis via caspase-3, 8, and 9, and lead to arrest at S/G2/M phases of cell cycle. The effects of combination therapy were more pronounced as compared to drug alone and they exert no toxicity to normal tissues (Mahbub et al. 2015). However, the combination of polyphenols and cytotoxic drugs were less effective in comparison to flavonoids. Phytochemicals along with their pharmacological action circumvent toxic effects of chemotherapeutic agents and also inhibit MDR pathways of cancer cells (Asensi et al. 2011).

6.5.2 Phytochemicals and NSAIDs

Combined use of phytochemicals along with miscellaneous synthetic drug like aspirin (ASP) followed by encapsulation in nanoparticles has been reported as a better strategy than using single molecule. Thakkar et al. (2013) evaluated the efficacy of low doses combination of curcumin, aspirin, and sulforaphane on pancreatic carcinoma. They reported that the combination effectively induced apoptosis and activated ERK1/2 signaling, thus inhibited cell growth (Thakkar et al. 2013). In another study, Guo et al. (2017) reported that the combination of aspirin and curcumin was more effective chemopreventive combination than single therapy. These results suggested that the low-doses combination therapy could be used to increase efficacy with minimal side effects. Hyaluronic acid anchored SLN for the

Table 6.1 Nanocarriers used for delivery of miscellaneous chemopreventive agents

Biomaterial	Carrier system	Encapsulated bioactive	Category	Cancer model	Reference
Calcium phosphate nanocores, enclosed in biodegradable polymers chitosan and alginate	Nanoparticles	Iron-saturated bovine lactoferrin (Fe-bLf) protein	Iron supplement	Colon cancer	Kanwar et al. (2012)
Poly(D,L-lactide-co-glycolide) (PLGA)	Nanoparticles	Meloxicam	COX-2 inhibitor	Colon adenocarcinoma	Şengel-Türk et al. (2012)
Iron oxide	Nanoparticles	Curcumin	Phytochemical	Breast cancer	Yallapu et al. (2012)
Solid lipid	Solid lipid nanoparticles	Raloxifen	Selective estrogen receptor modulator	Breast cancer	Battani et al. (2014)
Chitosan, Solid lipid	Chitosan-coated solid lipid nanoparticles	Ferulic acid and aspirin	Antioxidant and anti-inflammatory drug	Pancreatic cancer	Thakkar et al. (2015a, b)
Stearic acid, Compritol 888 ATO, and Tripalmitin	Solid lipid nanoparticles	Ibuprofen, Ibuprofen and sulphoraphane	Non-steroidal anti-inflammatory drug	Pancreatic cancer	Thakkar et al. (2015a, b)
Poly (lactic-co-glycolic acid)	Nanoparticles	Bromelain	Proteolytic enzyme	Skin carcinogenesis	Bhatnagar et al. (2015)
Silver	Nanoparticles	–	Anti-inflammatory and antimicrobial	Skin carcinogenesis	Arora et al. (2015)
Stearyl cetyl alcohol	Hyaluronic acid anchored solid lipid nanoparticles	Paclitaxel and ibuprofen	Anticancer and anti-inflammatory	Breast cancer	Tran et al. (2017)
Mesoporous silica	Nanoparticles	Curcumin	Phytochemical	Liver and cervical cancer	Elbialy et al. (2020)

combined delivery of paclitaxel and ibuprofen for chemoprevention is also reported (Tran et al. 2017).

Thakkar et al. (2015a, b) reported that solid lipid nanoparticles loaded with ferulic acid (FA) along with ASP successfully improved chemopreventive potential on tumor cell (Thakkar et al. 2015a, b). The FA + ASP treatment groups as compared to blank c-SLN vehicle control group showed reduced expression of marker of proliferation (Fig. 6.2). It indicates superiority of combined approach of using synthetic molecules with natural plant driven agents further synergized with use of

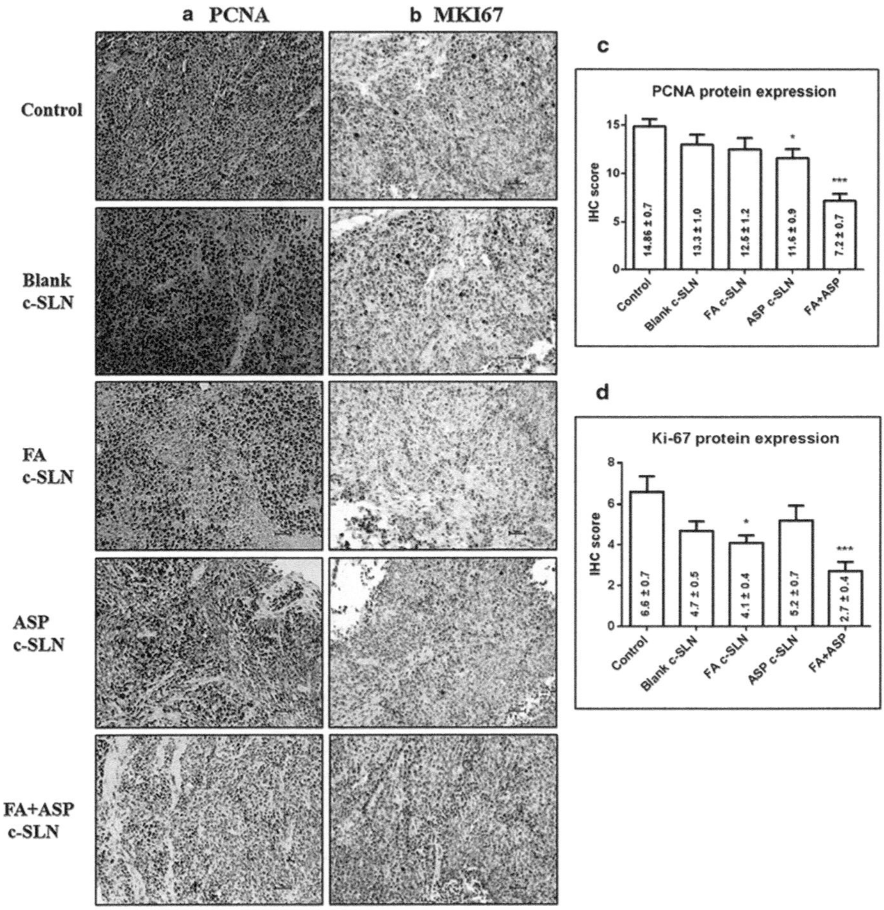

Fig. 6.2 Combination strategy used for synergistic chemopreventive action: effect of ferulic acid (FA) and aspirin (ASP) chemopreventive regimen on tumor cell proliferation and apoptosis. Immunohistochemical analysis (**a**, **b**) Representative images showing effect of modified FA and ASP c-SLN combination on PCNA and MKI67 (marker of proliferation Ki-67) expressions in pancreatic tissue. (**c**, **d**) A significant decrease in the PCNA and MKI67 expression was observed in modified FA + ASP treatment groups compared to blank c-SLN vehicle control group. (Adopted from Thakkar et al. 2015a, b)

nanoparticles in chemoprevention. However, an establishment of good relationship between combination uses of such synthetic molecules along with established natural agents and their chemopreventive effect is further needed.

Table 6.2 Theranostic delivery composites incorporating phytochemicals for cancer treatment and tumor imaging (Adopted from Lagoa et al. 2020)

Phytochemical	Delivery system	Therapeutic effects	Reference
Curcumin	Apoferritin loaded with gadolinium contrast agent	Induced cell death and reduced self-renewal in MDA-MB-231 and TUBO cells, and their derived cancer stem cells-enriched tumor spheres, and inhibition of breast cancer growth in vivo	Conti et al. (2016)
Gallic acid	NPs of zinc/aluminum-layered double hydroxide containing gadolinium and gold	Polyphenol and loaded carriers were more toxic to cancer HepG2 cells than to normal 3 T3 fibroblasts	Usman et al. (2017)
Protocatechuic acid	NPs of graphene oxide, gadolinium, and gold; NPs of magnesium/aluminum-layered (2018a, b) double hydroxide containing gadolinium and gold	Polyphenol and loaded carriers were more toxic to cancer HepG2 cells than to normal 3T3 fibroblasts	Usman et al. (2018a, b)
Quercetin	NPs of iron oxide coated with poly(citric acid)-PEG and folic acid	Empty carrier showed no effect, but the quercetin-loaded inhibited the growth of HeLa and MDA-MB-231 cells	Malekzadeh et al. (2017)
Tannic acid	NPs of self-assembled poly (propylene oxide) poloxamer Pluronic F-127 and tannic acid	For tumor imaging, NPs incorporating near-infrared fluorescent dye and Zr-89 accumulated in tumors, and showed good in vitro biocompatibility	Wang et al. (2018)
Resveratrol	Mannose-grafted albumin-QDs, co-delivery of pemetrexed	Internalization and cytotoxicity to breast cancer cells in vitro, and in vivo tumor growth inhibition, apoptosis promotion and angiogenesis inhibition superior to the free drugs. In addition to serum stability and hemato-compatibility of the system, no immunogenicity was detected in safety animal study	Zayed et al. (2019)

6.5.3 Combination of Diagnosis and Chemoprevention

Several studies that exploited nanotechnology and natural anticancer phytochemicals have been carried out for cancer therapy, chemoprevention, and diagnosis simultaneously (Table 6.2). Recently, Elbialy et al. developed PEG coated curcumin mesoporous silica nanoparticles as a theranostic agent (Elbialy et al. 2020). PEG-mesoporous silica-curcumin nanocarrier showed chemopreventive and cytotoxic agents along with as auto-fluorescences probe for diagnostic purpose. In a study, curcumin loaded self-fluorescent polymeric silica nanoparticles as theraonstics agent has been reported (Xu et al. 2018). Curcumin encapsulated magnetic nanoparticles were also developed for simultaneous drug delivery, drug targeting, and for real-time imaging of cancers (Yallapu et al. 2012). Such reports enlighten the base and future path for combined diagnosis and therapy using nanoparticles based chemopreventive agents.

6.6 Conclusion

Chemoprevention is a complex process and requires strategic planning to combat with the emerging and growing cancerous cells. A large number of chemopreventive agents are under evaluation either alone or in combination with synergizing agents. Pre-clinical reports are encouraging; however, a detailed analysis of each steps of their development with validation of the efficacy and safety both is further required. Nanoparticles loaded with combination of synthetic and phytochemicals for chemopreventive action via different mechanism should be tested in different clinical settings.

References

Arora S, Tyagi N, Bhardwaj A, Rusu L, Palanki R, Vig K, Singh SR, Singh AP, Palanki S, Miller ME, Carter JE (2015) Silver nanoparticles protect human keratinocytes against UVB radiation-induced DNA damage and apoptosis: potential for prevention of skin carcinogenesis. Nanomed Nanotechnol Biol Med 11(5):1265–1275

Asensi M, Ortega A, Mena S, Feddi F, Estrela JM (2011) Natural polyphenols in cancer therapy. Crit Rev Clin Lab Sci 48(5–6):197–216

Battani S, Pawar H, Suresh S (2014) Evaluation of oral bioavailability and anticancer potential of raloxifene solid lipid nanoparticles. J Nanosci Nanotechnol 14(8):5638–5645

Bartsch H, Ohshima H, Munoz N, Crespi M, Lu SH (1983) Measurement of endogenous nitrosation in humans: potential applications of a new method and initial results. Human 221:833–856

Bhatnagar P, Pant AB, Shukla Y, Chaudhari B, Kumar P, Gupta KC (2015) Bromelain nanoparticles protect against 7,12-dimethylbenz[a]anthracene induced skin carcinogenesis in mouse model. Eur J Pharma Biopharma 91:35–46

Bhol KC, Alroy J, Schechter PJ (2004) Anti-inflammatory effect of topical nanocrystalline silver cream on allergic contact dermatitis in a Guinea pig model. Clin Exp Dermatol Exp Dermatol 29 (3):282–287

Chan G, Boyle JO, Yang EK, Zhang F, Sacks PG, Shah JP, Edelstein D, Soslow RA, Koki AT, Woerner BM, Masferrer JL (1999) Cyclooxygenase-2 expression is up-regulated in squamous cell carcinoma of the head and neck. Cancer Res 59(5):991–994

Constantinou C, Papas A, Constantinou AI (2008) Vitamin E and cancer: an insight into the anticancer activities of vitamin E isomers and analogs. Int J Cancer 123(4):739–752

Conti L, Lanzardo S, Ruiu R, Cadenazzi M, Cavallo F et al (2016) L-ferritin targets breast cancer stem cells and delivers therapeutic and imaging agents. Oncotarget 7:66713–66727. https://doi.org/10.18632/oncotarget.10920

Ekstrand-Hammarström B, Österlund C, Lilliehöök B, Bucht A (2007) Vitamin E down-modulates mitogen-activated protein kinases, nuclear factor-κB and inflammatory responses in lung epithelial cells. Clin Exp Immunol 147(2):359–369

Elbialy NS, Aboushoushah SF, Sofi BF, Noorwali A (2020) Multifunctional curcumin-loaded mesoporous silica nanoparticles for cancer chemoprevention and therapy. Microporous Mesoporous Mater 291:109540

Gao F, Yuan Q, Gao L, Cai P, Zhu H, Liu R, Wang Y, Wei Y, Huang G, Liang J, Gao X (2014) Cytotoxicity and therapeutic effect of irinotecan combined with selenium nanoparticles. Biomaterials 35(31):8854–8866

Gengan RM, Anand K, Phulukdaree A, Chuturgoon A (2013) A549 lung cell line activity of biosynthesized silver nanoparticles using Albizia adianthifolia leaf. Colloids Surf B: Biointerfaces 105:87–91

Guo Y, Su ZY, Zhang C, Gaspar JM, Wang R, Hart RP, Verzi MP, Kong ANT (2017) Mechanisms of colitis-accelerated colon carcinogenesis and its prevention with the combination of aspirin and curcumin: transcriptomic analysis using RNA-seq. Biochem Pharmacol 135:22–34

Gurunathan S, Lee KJ, Kalishwaralal K, Sheikpranbabu S, Vaidyanathan R, Eom SH (2009) Antiangiogenic properties of silver nanoparticles. Biomaterials 30(31):6341–6350

Hsin YH, Chen CF, Huang S, Shih TS, Lai PS, Chueh PJ (2008) The apoptotic effect of nanosilver is mediated by a ROS-and JNK-dependent mechanism involving the mitochondrial pathway in NIH3T3 cells. Toxicol Lett 179(3):130–139

Ip M, Lui SL, Poon VK, Lung I, Burd A (2006) Antimicrobial activities of silver dressings: an in vitro comparison. J Med Microbiol 55(1):59–63

Jain J, Arora S, Rajwade JM, Omray P, Khandelwal S, Paknikar KM (2009) Silver nanoparticles in therapeutics: development of an antimicrobial gel formulation for topical use. Mol Pharm 6 (5):1388–1401

Jeyaraj M, Rajesh M, Arun R, MubarakAli D, Sathishkumar G, Sivanandhan G, Dev GK, Manickavasagam M, Premkumar K, Thajuddin N, Ganapathi A (2013a) An investigation on the cytotoxicity and caspase-mediated apoptotic effect of biologically synthesized silver nanoparticles using *Podophyllum hexandrum* on human cervical carcinoma cells. Colloids Surf B: Biointerfaces 102:708–717

Jeyaraj M, Sathishkumar G, Sivanandhan G, MubarakAli D, Rajesh M, Arun R, Kapildev G, Manickavasagam M, Thajuddin N, Premkumar K, Ganapathi A (2013b) Biogenic silver nanoparticles for cancer treatment: an experimental report. Colloids Surf B: Biointerfaces 106:86–92

Jones SA, Bowler PG, Walker M, Parsons D (2004) Controlling wound bioburden with a novel silver-containing Hydrofiber® dressing. Wound Repair Regen 12(3):288–294

Kanwar JR, Mahidhara G, Kanwar RK (2012) Novel alginate-enclosed chitosan–calcium phosphate-loaded iron-saturated bovine lactoferrin nanocarriers for oral delivery in colon cancer therapy. Nanomedicine 7(10):1521–1550

Kokura S, Handa O, Takagi T, Ishikawa T, Naito Y, Yoshikawa T (2010) Silver nanoparticles as a safe preservative for use in cosmetics. Nanomed Nanotechnol Biol Med 6(4):570–574

Koornstra JJ, Rijcken FE, Oldenhuis CN, Zwart N, Van Der Sluis T, Hollema H, Elisabeth GE, Keller JJ, Offerhaus JA, Giardiello FM, Kleibeuker JH (2005) Sulindac inhibits β-catenin expression in normal-appearing colon of hereditary nonpolyposis colorectal cancer and familial adenomatous polyposis patients. Cancer Epidemiol Prev Biomarkers 14(7):1608–1612

Kristal AR (2004) Vitamin A, retinoids and carotenoids as chemopreventive agents for prostate cancer. J Urol 171(2):S54–S58
Lagoa R, Silva J, Rodrigues JR, Bishayee A (2020) Advances in phytochemical delivery systems for improved anticancer activity. Biotechnol Adv 38:107382
Lazovich D, Choi K, Vogel RI (2012) Time to get serious about skin cancer prevention. Cancer Epidemiol Prev Biomarkers 21(11):1893–1901
Liao W, Zhang R, Dong C, Yu Z, Ren J (2016) Novel walnut peptide–selenium hybrids with enhanced anticancer synergism: facile synthesis and mechanistic investigation of anticancer activity. Int J Nanomed 11:1305
Lotan Y, Xu XC, Shalev M, Lotan R, Williams R, Wheeler TM, Thompson TC, Kadmon D (2000) Differential expression of nuclear retinoid receptors in normal and malignant prostates. J Clin Oncol 18(1):116–116
Lowe SW, Cepero E, Evan G (2004) Intrinsic tumour suppression. Nature 432(7015):307–315
Mahbub AA, Le Maitre CL, Haywood-Small SL, Cross NA, Jordan-Mahy N (2015) Polyphenols act synergistically with doxorubicin and etoposide in leukaemia cell lines. Cell Death Discov 1 (1):1–12
Malekzadeh A, Ramazani A, Rezaei S, Niknejad H (2017) Design and construction of multifunctional hyperbranched polymers coated magnetite nanoparticles for both targeting magnetic resonance imaging and cancer therapy. J Colloid Interface Sci 490:64–73. https://doi.org/10.1016/j.jcis.2016.11.014
Marnett LJ (1994) Generation of mutagens during arachidonic acid metabolism. Cancer Metastasis Rev 13(3–4):303–308
Menon S, Shanmugam V (2020) Chemopreventive mechanism of action by oxidative stress and toxicity induced surface decorated selenium nanoparticles. J Trace Elem Med Biol 62:126549
Munger MA, Radwanski P, Hadlock GC, Stoddard G, Shaaban A, Falconer J, Grainger DW, Deering-Rice CE (2014) In vivo human time-exposure study of orally dosed commercial silver nanoparticles. Nanomed Nanotechnol Biol Med 10(1):1–9
Narayanan BA, Narayanan NK, Desai D, Pittman B, Reddy BS (2004) Effects of a combination of docosahexaenoic acid and 1, 4-phenylene bis (methylene) selenocyanate on cyclooxygenase 2, inducible nitric oxide synthase and β-catenin pathways in colon cancer cells. Carcinogenesis 25(12):2443–2449
Pangburn HA, Ahnen DJ, Rice PL (2010) Sulindac metabolites induce proteosomal and lysosomal degradation of the epidermal growth factor receptor. Cancer Prev Res 3(4):560–572
Peto RRJM, Doll R, Buckley JD, Sporn MB (1981) Can dietary beta-carotene materially reduce human cancer rates? Nature 290(5803):201–208
Rao CV, Reddy BS (2004) NSAIDs and chemoprevention. Curr Cancer Drug Targets 4(1):29–42
Reed JC (2006) Drug insight: cancer therapy strategies based on restoration of endogenous cell death mechanisms. Nat Clin Pract Oncol 3(7):388–398
Şengel-Türk CT, Hasçiçek C, Dogan AL, Esendagli G, Guc D, Gönül N (2012) Preparation and evaluation of meloxicam-loaded PLGA nanoparticles on HT-29 human colon adenocarcinoma cells. Drug Dev Ind Pharm 38(9):1107–1116
Seo SK, Lee HC, Woo SH, Jin HO, Yoo DH, Lee SJ, An S, Choe TB, Park MJ, Hong SI, Park IC (2007) Sulindac-derived reactive oxygen species induce apoptosis of human multiple myeloma cells via p38 mitogen activated protein kinase-induced mitochondrial dysfunction. Apoptosis 12 (1):195
Smolarek AK, Suh N (2011) Chemopreventive activity of vitamin E in breast cancer: a focus on γ-and δ-tocopherol. Nutrients 3(11):962–986
Smolarek AK, So JY, Burgess B, Kong ANT, Reuhl K, Lin Y, Shih WJ, Li G, Lee MJ, Chen YK, Yang CS (2012) Dietary administration of δ-and γ-tocopherol inhibits tumorigenesis in the animal model of estrogen receptor–positive, but not HER-2 breast cancer. Cancer Prev Res 5 (11):1310–1320
Soh JW, Weinstein IB (2003) Role of COX-independent targets of NSAIDs and related compounds in cancer prevention and treatment. Prog Exp Tumor Res 37:261–283

Subbaramaiah K, Dannenberg AJ (2003) Cyclooxygenase 2: a molecular target for cancer prevention and treatment. Trends Pharmacol Sci 24(2):96–102
Summerfield FW, Tappel AL (1984) Vitamin E protects against methyl ethyl ketone peroxide-induced peroxidative damage to rat brain DNA. Mutat Res Fundam Mol Mech Mutagen 126 (2):113–120
Takada Y, Bhardwaj A, Potdar P, Aggarwal BB (2004) Nonsteroidal anti-inflammatory agents differ in their ability to suppress NF-κ B activation, inhibition of expression of cyclooxygenase-2 and cyclin D1, and abrogation of tumor cell proliferation. Oncogene 23(57):9247–9258
Tanaka T, Shnimizu M, Moriwaki H (2012) Cancer chemoprevention by carotenoids. Molecules 17 (3):3202–3242
Thakkar A, Sutaria D, Grandhi BK, Wang J, Prabhu S (2013) The molecular mechanism of action of aspirin, curcumin and sulforaphane combinations in the chemoprevention of pancreatic cancer. Oncol Rep 29(4):1671–1677
Thakkar A, Chenreddy S, Wang J, Prabhu S (2015a) Ferulic acid combined with aspirin demonstrates chemopreventive potential towards pancreatic cancer when delivered using chitosan-coated solid-lipid nanoparticles. Cell Biosci 5:1–14
Thakkar A, Chenreddy S, Wang J, Prabhu S (2015b) Evaluation of ibuprofen loaded solid lipid nanoparticles and its combination regimens for pancreatic cancer chemoprevention. Int J Oncol 46(4):1827–1834
Tozawa K, Sakurada S, Kohri K, Okamoto T (1995) Effects of anti-nuclear factor κ B reagents in blocking adhesion of human cancer cells to vascular endothelial cells. Cancer Res 55 (18):4162–4167
Tran BN, Nguyen HT, Kim JO, Yong CS, Nguyen CN (2017) Combination of a chemopreventive agent and paclitaxel in CD44-targeted hybrid nanoparticles for breast cancer treatment. Arch Pharm Res 40(12):1420–1432
Trump DL (1994) Retinoids in bladder, testis and prostate cancer: epidemiologic, pre-clinical and clinical observations. Leukemia 8:S50–S54
Tsujii M, DuBois RN (1995) Alterations in cellular adhesion and apoptosis in epithelial cells overexpressing prostaglandin endoperoxide synthase 2. Cell 83(3):493–501
Usman M, Hussein M, Fakurazi S, Masarudin M, Ahmad Saad F (2017) Gadoliniumdoped gallic acid-zinc/aluminium-layered double hydroxide/gold theranostic nanoparticles for a bimodal magnetic resonance imaging and drug delivery system. Nanomaterials 7:244. https://doi.org/10.3390/nano7090244
Usman M, Hussein M, Kura A, Fakurazi S, Masarudin M, Ahmad Saad F (2018a) Graphene oxide as a nanocarrier for a theranostics delivery system of protocatechuic acid and gadolinium/gold nanoparticles. Molecules 23:500. https://doi.org/10.3390/molecules23020500
Usman M, Hussein M, Kura A, Fakurazi S, Masarudin M, Saad F (2018b) Synthesis and characterization of protocatechuic acid-loaded gadolinium-layered double hydroxide and gold nanocomposite for theranostic application. Appl Nanosci 8:973–986. https://doi.org/10.1007/s13204-018-0752-6
Wang X, Sun K, Tan Y, Wu S, Zhang J (2014) Efficacy and safety of selenium nanoparticles administered intraperitoneally for the prevention of growth of cancer cells in the peritoneal cavity. Free Radic Biol Med 72:1–10
Wang X, Yan J, Pan D, Yang R, Wang L et al (2018) Polyphenol–poloxamer selfassembled supramolecular nanoparticles for tumor NIRF/PET imaging. Adv Healthc Mater 7:e1701505. https://doi.org/10.1002/adhm.201701505
Xu X, Lü S, Wu C, Wang Z, Feng C, Wen N, Liu M, Zhang X, Liu Z, Liu Y, Ren C (2018) Curcumin polymer coated, self-fluorescent and stimuli-responsive multifunctional mesoporous silica nanoparticles for drug delivery. Microporous Mesoporous Mater 271:234–242
Yallapu MM, Othman SF, Curtis ET, Bauer NA, Chauhan N, Kumar D, Jaggi M, Chauhan SC (2012) Curcumin-loaded magnetic nanoparticles for breast cancer therapeutics and imaging applications. Int J Nanomed 7:1761

Zayed D, Ebrahim S, Helmy M, Khattab S, Bahey-El-Din M et al (2019) Combining hydrophilic chemotherapy and hydrophobic phytotherapy via tumor-targeted albumin–QDs nano-hybrids: covalent coupling and phospholipid complexation approaches. J Nanobiotechnol 17:7. https://doi.org/10.1186/s12951-019-0445-7

Zheng S, Li X, Zhang Y, Xie Q, Wong YS, Zheng W, Chen T (2012) PEG-nanolized ultrasmall selenium nanoparticles overcome drug resistance in hepatocellular carcinoma HepG2 cells through induction of mitochondria dysfunction. Int J Nanomed 7:3939

Zhou Y, Xu M, Liu Y, Bai Y, Deng Y, Liu J, Chen L (2016) Green synthesis of Se/Ru alloy nanoparticles using gallic acid and evaluation of their anti-invasive effects in HeLa cells. Colloids Surf B: Biointerfaces 144:118–124

Quality Control, Scale-Up, and Regulatory Aspects of Herbal Nanomedicine

7

Abstract

Herbal nanomedicine is one of the fastest growing research areas. The regained interests in the cancer treatment and chemoprevention using herbal chemicals have turned the coin with the help of nanotechnology. Regulatory concerns related to nanotechnology and herbal formulations are being used at present to assess the essential characteristics of herbal nanomedicines. In this chapter, we have summarized all the key points and progress of the herbal nanomedicine including discussion on quality control of nanoparticles, toxicity consideration of nanomedicines, and assessment and importance of efficacy, safety, and quality control of herbal medicines.

Keywords

Herbal product · Quality assurance · Product development · Regulatory guidelines · Nanotechnology · Chemoprevention

7.1 Introduction

Modern era of medicine is based on technology including 3D printing, nanotechnology, and personalized medicine. Nanomedicine is a well-established concept particularly for rare disease like cancer and cancer prevention. Globally, market for nanomedicine is increasing day by day for treatment, diagnosis, and prevention of ailments (Marques et al. 2019). As the name indicates, nanomedicine is referring to nanotechnology and medicine both together for precise and controlled medication for the patients' compliances. At some places, nanotechnology along with pharmaceuticals is known as nanopharmaceuticals (Agarwal et al. 2018). The convergence and upcoming of nanotechnology-related product in the medicine have arisen some challenges towards regulatory agencies as along with said benefit

R. Paliwal, S. R. Paliwal, *Advances in Nanochemoprevention*,
https://doi.org/10.1007/978-981-15-9692-6_7

of these tiny particles, some potential risk and hazards are also associated with them. Nanoparticles may impose these complications to human health as well as environments (Wacker et al. 2016).

Technically, nanotechnology deals with the materials in the size range of 0.1 nm to 100 nm. At such small size, the properties of the materials drastically change than that of their bulk part like electrical conductance, optical properties, chemical reactivity, magnetism, and physical strength (Sarwal et al. 2018). In general, any substance between 1 nm and 100 nm is considered nanomaterials. Apart from the benefits from nanomaterials, the limiting factors like possible toxicity, scale-up issues, and regulatory guidelines require relevant characterization and assays (Cui et al. 2019; Lameijer et al. 2018; Haque et al. 2016; Ermolin et al. 2018; Hou et al. 2018; Accomasso et al. 2018).

Natural medication of the diseases is quite popular in the society since ancient time and gained trust being time tested. A large number of plants and their parts are being used to treat variety of the diseases all over the world; however, the scientific documentation is comparatively lacking than it is being practiced. In several countries, herbals drugs are still semi-regulated and lacking registered practices and strict supervision of health authorities and therefore safety concern is not addressed properly. Herbal medicine is generally referred as safe in comparison to synthetic drug molecules (Bhatt 2016); however, after encapsulation of phytodrugs into nanoparticles the issues associated with nanoparticles are to be considered. In order to regulate and document, USFDA has evaluated adverse effect of plants and other dietary supplements. It is also observed that for a large number of herbal drugs, the efficacy is not established and no quality assurance. World Health Organization's (WHO) as a part of Traditional Medicine (TM) Strategy 2014–2023 focused on promoting the safety, efficacy, and quality of TM by growing the information base and providing guidance on regulatory and quality assurance standards (www.who.int).

Major regulatory agencies worldwide for nanomaterials/nanopharmaceuticals

Country	Regulatory agencies	Website
USA	Food and Drug Administration	www.fda.gov
Europe	European Commission	https://ec.europa.eu
Japan	Pharmaceuticals and Medical Devices Agency	https://www.pmda.go.jp
India	Department of Science and Technology	https://dst.gov.in
China	Ministry of Science and Technology (MOST)	http://www.most.gov.cn
Canada	Health Canada, Canada	https://www.canada.ca/en/health-canada.html
Switzerland	Swissmedic	https://www.swissmedic.ch/swissmedic/en/home.html
Netherlands	National Institute for Public Health and the Environment	https://www.rivm.nl/en
Taiwan	Centre for Drug Evaluations	https://www.cde.org.tw/eng/

(continued)

Brazil	Brazilian Health Regulatory Agency (ANVISA)	http://portal.anvisa.gov.br/english
Korea	Ministry for Food and Drug Administration	https://www.mfds.go.kr/eng/index.do

At the moment, specific guidelines for nanomaterials and nanomedicine are under process. The regulatory agencies like USFDA and EMA are in connection with the established researchers and industry stakeholders for developing guidelines documents (Flühmann et al. 2019). Further, it is expected that components of nanotechnology-based new drug applications will be soon included into the specific regulations throughout the world (Amenta et al. 2015; Musazzi et al. 2017).

In 2017, WHO has developed guidelines with recommendations for protecting workers from the potential risks of engineered nanomaterials (ENMs). These guidelines include a definition for ENMs and specify equipment as well as preventive measures to reduce occupational exposure to ENMs (Rodríguez-Ibarra et al. 2020). In the United States of America, USFDA regulates safety and efficacy of almost all kind of medical products including drugs, biologics, and devices via federal legislation and agency regulations and policy (Paradise 2019). Nanomaterial based therapeutics challenge the conventional regulatory framework that results due to their complex combinatorial characteristics with diversified mechanism of their action therapeutically. About more than 100 nanomedicine products have been approved by USFDA and many more are in the pipelines. A systematic regulation is required to assess the safety, efficacy, reproducibility, bioequivalence, and scale-up possibilities of these nanosystems.

Recently, the guidelines for evaluation of nanopharmaceuticals in India have been launched in October 2019. Similarly, many countries are also working in order to reach the legal solutions to deal with nanotechnology and their complications. Though, at present diversified legislations are available on different sectors like environment, occupational health, product labeling and liability, healthcare, chemical, and cosmetic legislation (Wacker et al. 2016; Bremer-Hoffmann et al. 2018).

Not only in medicine and health but also in agriculture and food sectors, specific guidelines are being developed thought the world. EU has already included nano-specific provision in the existing legislations. At present, the world's largest economies, the United States of America (USA), China, EU, Canada, Switzerland, Japan, China, and India are setting global standards in the definition (see Table 7.1), characterization, and safety assessment of nanomaterials (Marques et al. 2019; Sarwal et al. 2018).

7.2 Nanotechnology in Medicine

Nanotechnologies contribute to almost every field of science, including physics, materials science, chemistry, biology, computer science, and engineering. Remarkably, in recent years nanotechnologies have been applied to human health with

Table 7.1 Definitions applied to nanotechnology-related products in the different countries

Country	Definition
USA	A material or end product is engineered to have at least one external dimension, or an internal or surface structure, in the nanoscale range(approximately 1 nm to 100 nm);—whether a material or end product is engineered to exhibit properties or phenomena, including physical or chemical properties or biological effects, that are attributable to its dimension(s), even if these dimensions fall outside the nanoscale range, up to 1 μm (1000 nm)
European Union	A natural, incidental, or manufactured material containing particles, in an unbound state or as an aggregate or as an agglomerate and where, for 50% or more of the particles in the number size distribution, one or more external dimensions are in the size range of 1–100 nm
China	Nanomaterial is a material which has a structure in the three-dimensional space in at least one dimension in the nanometer scale (from 1 nm to100 nm) range of geometric dimensions, or constituted by the nanostructure unit and a material with special properties
Canada	Any manufactured substance or product and any component material, ingredient, device, or structure to be nanomaterial if: (a) it is at or within the nanoscale (1–100 nm) in at least one external dimension, or has internal or surface structure at the nanoscale; or (b) it is smaller or larger than the nanoscale in all dimensions and exhibits one or more nanoscale properties/phenomena
Switzerland	The particles that have at least one nanoscale dimension (1–1000 nm) plus a function and/or mode of action based on nanotechnology characteristics
Japan	No definitions as such. But, as in the reflection paper on block copolymer micelles, there is a sentence as follows: "As block copolymer micelle products are of nanoscale size, contain more than one component, and are purposely designed for specific clinical applications they may be considered as nanomedicines
India	The nanomaterial is generally defined as material having particle size in the range of 1–100 nm in at least one dimension. However, if a material exhibits physical, chemical, or biological phenomenon or activity which are attributable to its dimension beyond nanoscale range up to 1000 nm, the material should also be considered as nanomaterial

promising results, particularly in the field of cancer treatment (Bayda et al. 2020). Nanomedicine expanded multiple areas, including drug delivery, vaccine development, antibacterial, diagnosis and imaging tools, wearable devices, implants, high-throughput screening platforms, etc., using biological, nonbiological, biomimetic, or hybrid materials (Soares et al. 2018). Figure 7.1 shows the progression of nanoscience and technology in different domains.

7.3 Major Regulatory Concerns for Nanoparticles in Therapeutics

In 2006, Organization for Economic Co-operation and Development (OECD) launched a program of the work on hazard, exposure, and risk assessment of nanomaterials. OECD started working on regulations for manufactured

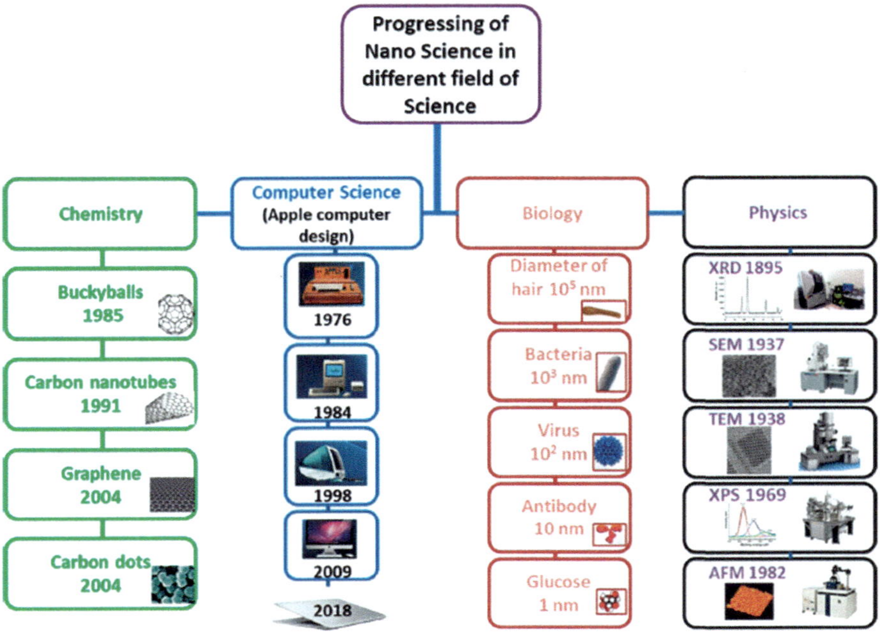

Fig. 7.1 Progress of nanotechnology in different field of science (Adopted from Bayda et al. 2020)

nanomaterials with the International Organization for Standardization (ISO) to provide scientific advice for the safety use of nanomaterials that includes the respective physicochemical characterization and the metrology (Soares et al. 2018). However, till date there is not a comprehensive and effective list of minimum parameters for nanomaterial characterization and it is ever expanding and changing with new advancement in the science and technology day by day. The following characteristics should be a starting point to the characterization: particle size, shape and size distribution, aggregation and agglomeration state, crystal structure, specific surface area, porosity, chemical composition, surface chemistry, charge, photocatalytic activity, zeta potential, water solubility, dissolution rate/kinetics, and dustiness.

7.4 Quality Control of Nanomaterials

Nanomaterials need strict quality control for claiming said properties particularly in health and medicine. These essential parameters are highly critical in nature for the desired performance and therefore require suitable characterization techniques. In general, the size in minimum one direction of the particles should be in between 1 and 100 nm with more than 50% number size distribution (Soares et al. 2018). Some of the structure like fullerenes, graphenes, and single-wall carbon nanotubes may have size less than 1 nm and considered as nanomaterials. Since 2017, it is now essential for nano-sized material to follow guidelines of Toxic Substances Control Act (TSCA) according to the Environmental Protection Agency (EPA) in the USA (Rodríguez-Ibarra et al. 2020). Furthermore, the information like specific chemical identity, production volume, manufacturing technique, processing, use, exposure and release, available health and safety data are required related to any nanomaterial (Rodríguez-Ibarra et al. 2020). The handling of nanomaterials in the research labs is also crucial to ensure the safety of the personnel involved in the work (Stuttgen et al. 2019). It has been observed that regulations for toxic materials and carcinogens are not clear with regard to nanomaterials and need clear guidelines in future. According to the "Nanomaterial Registry" of National Institute of Health (NIH), 12 characteristics of nanomaterials are required as minimum information. These include particle size and distribution, composition, shape, surface area and charge, purity, agglomeration/aggregation state, solubility, surface reactivity, chemistry, and stability. Some of these characteristics are described.

7.4.1 Particle Size

Particle size is the first characteristics to define the particle as "nanoparticles." As discussed, the usual particle size range is 1 nm–100 nm. There is no limit of this range and depends upon the desired and acceptable features in the particles required and its applications. In medicine, a nanomaterial is considered nanomedicine if it supports and maintains desired physicochemical, biological, and therapeutic performance. Few nanomaterials contain their one dimension in nanosize (film or coating), some have two dimensions in nanosize (tube, fiber, or a wire); while some have all three dimensions in the range (quantum dots or hollow spheres). Though FDA has not set any definition for the term "nanotechnology" or "Nanomaterial" or "Nano-scale" or "Nanoparticles" or any related terminology; however, any engineered material having at least one dimension in the above-mentioned range is considered as nanomaterial (Rodríguez-Ibarra et al. 2020). Photon correlation spectroscopy or dynamic light scattering technique is commonly employed for determination of particle size.

7.4.2 Particle Size Distribution

Particle size distribution is an important parameter as it describes the variation in the sizes of the particles in the dispersion. It clearly demonstrates the extent of polydispersity of the nanoparticles compositions (Soares et al. 2018). The light or X-ray illumination scattering techniques are used on the basis of mass or electron distribution of the nanoparticles (Modena et al. 2019).

7.4.3 Particle Shape

Nanoparticles having similar composition or the dimensions may have drastically different behaviors. Nanoparticles shape is also important in therapeutics for some functional characteristics like surface binding capabilities, drug release, cellular penetration, opt, and aggregation behavior. A large variety of shapes geometric as well as irregular are reported in the literature. High-resolution microscopy techniques like electron microscopy and scanning probe microscopy, etc., are used to observe the shape of nanoparticles under investigation as they provide detailed information on shape, surface texture, artifacts, aggregation, etc. (Modena et al. 2019). A limitation of electron microscopy is that it provides only two-dimensional projections of particle shape.

7.4.4 Particle Density

The effective density and hydrodynamic diameter of nanoparticles both are required for their desired transport, fate, interaction, and internalization through cell surface proteins (Watson et al. 2016). It has been reported that effective density for non-buoyant nanoparticles is 1.578 gcm^{-3}. Further, low density may lead to inaccurate dose response in in vitro cell culture studies making them buoyant nanoparticles and therefore poor assessment of efficacy and cytotoxicity. Henceforth, coating of a polymer may alter the density of the nanoparticles and hence their performance as well (Suk et al. 2016).

7.5 Toxicity Consideration of Nanomedicine

The market is receiving nanomedicine based product continuously and products are being commercialized (at present more than 100 USFDA approved product). Many more products are in the process of the approval including generics as well (Umezawa et al. 2017). However, as mentioned, the adverse effect and toxicity needs to be evaluated before their claims for therapeutics in healthcare. A long-term basis studies are required for safety, efficacy, and compliances of nanomedicine based products. More advanced functional analysis of these products at genomics, proteomics, transcriptomics, and metabolics may provide higher level of safety data and information. The exposure of the nanoparticles to vital healthy organs and their subsequent effects should be studied extensively and reported. For examples, Umezawa et al. 2017 studied developmental toxicity of nanoparticles on the brain. The dose- and size-dependency of transplacental nanoparticle transfer to the fetus were reported. It further indicates the importance of understanding both the mechanism of direct effect of nanoparticles transferred to the fetus and offspring and the indirect effect mediated by induction of oxidative stress and inflammation in the pregnant body. Locomotor activity, learning and memory, motor coordination, and social behavior were reported as potential neurobehavioral targets of maternal nanoparticle exposure. In another study, the exposure of human endothelial cells (ECs) to nanoparticles (NPs) could lead to cytotoxicity, genotoxicity, endothelial activation, and impaired NO signaling (Cao 2018). Authors suggested that oxidative stress and inflammation induced by NPs were the major mechanisms associated with the toxicity of NPs to ECs. A three-tier model was further proposed to explain the association between NP induced oxidative stress and toxicity. Another major organ environmentally exposed with nanoparticles is skin. It serves as a major body organ that acts as the first-line barrier between the internal organs and external environment (Hashempour et al. 2019). Author elaborated the clinical side effects of nanoparticles following topical admonition, including skin inflammation, skin cancer, and genetic toxicity. Although the inhalation and ingestion of nanoparticles are more dangerous compared with skin exposure, there are noteworthy information gaps in skin exposure to nanoparticles that need much attention. Silicon-based materials and their oxides are widely used in drug delivery, dietary supplements, implants, and dental fillers (Chen et al. 2018). Silica nanoparticles (SiNPs) interact with immunocompetent cells and induce immunotoxicity.

7.6 Efficacy, Safety, and Quality Control of Herbal Medicines

Herbal medicinal product including nanophytopharmaceuticals should fulfill all the requirements for quality, safety, and efficacy like other medicinal products (Steinhoff 2019). Herbal products require specific production and quality control features like essential purity, identification, assays, heavy metal analysis due to their complex characters. The legal provisions of herbal products are parts of pharmacopeial monographs and documented by various regulatory agencies. Pharmacopeia legal provisions include quantification and identification of potential residues (pesticides, heavy metal, elemental impurities, mycotoxins, and microorganism) and containments in either herbal substance or preparations used for herbal medicinal product development (Steinhoff 2019). Figure 7.2 shows some of the European guidelines for testing herbal medicinal plants.

Regulations for herbal formulations are different for different countries. However, WHO has set certain minimum standards for evaluation and quality control of herbal drugs (www.who.int/medicinedocs/pdf/whozip57e/whozip57e.pdf). Table 7.2 lists these parameters for quick outlook and are compiled from Surekha et al. 2016.

The USFDA, Botanical Drug Development Guidance for Industry, described the appropriate development plans for botanical drugs for submission along with new drug applications (NDAs) (Modena et al. 2019). Figure 7.3 highlights the development steps for herbal drugs products as per USFDA guidelines for industry. The process is divided into two steps for easy understanding to readers.

Apart from general regulatory requirements for an NDA application, additional information is also required for botanical drugs such as nonclinical pharmacology/ toxicology studies, clinical evidence of efficacy and safety. Such special requirements to ensure safety and quality of botanicals are essential prerequisite

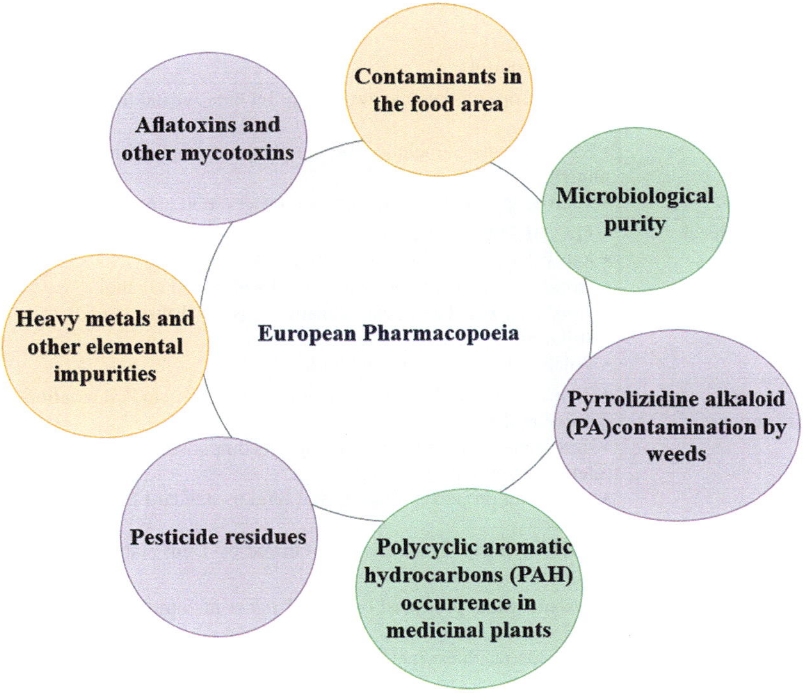

Fig. 7.2 Pharmacopeia test for assessment of herbal medicines as per European guidelines

for product development. Recently, the risk of hepatotoxicity has been linked with few herbal products in the USA and this draws attention of regulatory community. Nowadays, all dietary supplements including botanicals that are sold domestically are categorized as special category of foods by USFDA (Avigan et al. 2016). The need of transparent scientific quality standards has been advocated looking into global supply chains of these dietary supplements (Thakkar et al. 2020). Traditional medicines of India and their quality control test as mentioned by Regulatory Agency of India, AYUSH, have been described by Mukherjee et al. 2016.

Quality consistencies of herbal drug products should be ensured to achieve the claimed therapeutic efficacy as they may change due to physicochemical complexities and variability (Kim et al. 2019). The overall quality and performance depend upon several factors like collection site, weather, age of the medicinal plants, species, harvest season and time, developmental stage, temperature, and humidity on plant metabolite production, etc. (Pferschy-Wenzig and Bauer 2015). For more detail of the guidelines and regulations, readers are suggested to read article published recently by Kim et al. 2019 where testing parameters and methods for finished herbal products are detailed in the guidelines and regulations issued by five global authorities and 15 countries. Nowadays, DNA barcoding along with chromatographic fingerprint is being used for the assessment and quality control of

Table 7.2 WHO Parameters for quality control of herbal drugs (compiled from Surekha et al. 2016)

Parameters	Quality Control
Botanical parameters	• Sensory or organoleptic evaluation: Includes visual macroscopy/touch/odor/taste • Foreign matters: Includes foreign plants, foreign animals, foreign minerals • Microscopy: Includes histological observation and measurements
Physicochemical properties	• TLC/HPTLC fingerprint • Ash values: Total, acid-insoluble, water-soluble • Extractive values; in hot water, cold water, and ethanol • Moisture content and volatile matter: Loss on drying (LOD), azeotropic distillation • Volatile oils: By steam distillation
Pharmacological parameters	• Bitterness value: Unit equivalent bitterness of standard solution of quinine hydrochloride • Hemolytic property: On ox blood by comparison with standard reference solution of saponin • Astringent property: Tannins that bind to standard Freiberg Hide powder • Foaming index: Foam height produced by 1 gm material under specified conditions
Toxicological parameters	• Arsenic: Stain produced on HgBr2 paper in comparison to standard stain • Pesticide residues: Includes total organic chloride and total organic phosphorous • Heavy metals: Like cadmium and lead • Microbial contamination: Total viable aerobic count of pathogens: *Enterobacteria, E. coli, Salmonella, P. aeruginosa, S. aureus* • Aflatoxins: By TLC using aflatoxins (B1, B2, G1, and G2) • Radioactive contamination

herbal medicinal products to evaluate the authenticity and quality consistency (Mohammed Abubakar et al. 2017).

7.7 Conclusion

Nanomedicine or nanoparticles for therapeutics offer advantages and challenges both in terms of quality control and regulatory aspects. Loading of phytopharmaceuticals into nanoparticles renders dual regulations of both nanotechnology and herbal formulations in case of plant extracts. Since, nanoparticles alter both physicochemical and pharmacokinetic parameters, therefore require special attention in term of their biodistribution, degradation, and toxicity towards safe and vital organs. Reproduction and scale-up possibilities are required characteristics to be successful as commercial products. Globally, a large number of guidelines are available and if not are being made available by the regulatory agencies. However, till date no specific guidelines for nanochemopreventive agents are available for

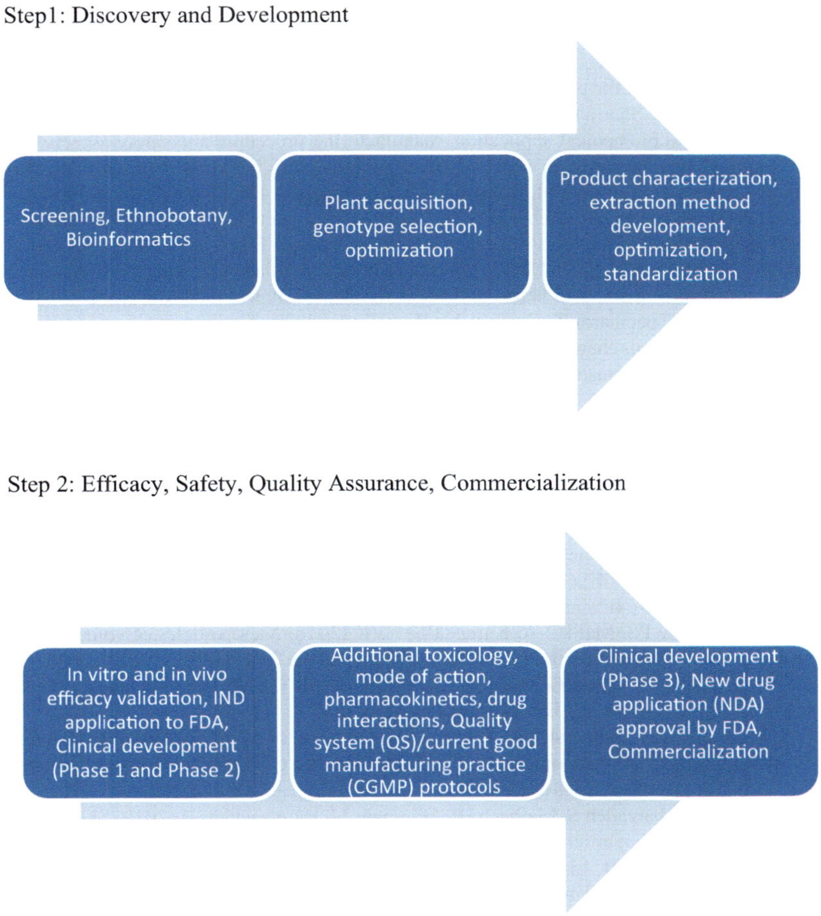

Fig. 7.3 Essential steps involved in the process of development of botanical drug products

immediate referral. In future, it is expected that increasing demands of such commercial products will demand to generate the data for developing such guidelines. Till date, the safety and quality control guidelines of WHO, OECD, ICH, and regulatory agencies of different countries for both nanotechnology and herbal formulation altogether should be considered as the milestone in this process of nanochemopreventive product development. Since developing quality product is a continuous process, more reports are required to generate the preliminary data related to such products.

References

Accomasso L, Cristallini C, Giachino C (2018) Risk assessment and risk minimization in nanomedicine: a need for predictive, alternative, and 3Rs strategies. Front Pharmacol 9:228

Agarwal V, Bajpai M, Sharma A (2018) Patented and approval scenario of nanopharmaceuticals with relevancy to biomedical application, manufacturing procedure and safety aspects. Recent Pat Drug Delivery Formulation 12(1):40–52

Amenta V, Aschberger K, Arena M, Bouwmeester H, Moniz FB, Brandhoff P, Gottardo S, Marvin HJ, Mech A, Pesudo LQ, Rauscher H (2015) Regulatory aspects of nanotechnology in the Agri/feed/food sector in EU and non-EU countries. Regul Toxicol Pharmacol 73(1):463–476

Avigan MI, Mozersky RP, Seeff LB (2016) Scientific and regulatory perspectives in herbal and dietary supplement associated hepatotoxicity in the United States. Int J Mol Sci 17(3):331

Bayda S, Adeel M, Tuccinardi T, Cordani M, Rizzolio F (2020) The history of nanoscience and nanotechnology: from chemical–physical applications to nanomedicine. Molecules 25(1):112

Bhatt A (2016) Phytopharmaceuticals: a new drug class regulated in India. Perspect Clin Res 7 (2):59

Bremer-Hoffmann S, Halamoda-Kenzaoui B, Borgos SE (2018) Identification of regulatory needs for nanomedicines. J Interdiscip Nanomed 3(1):4–15

Cao Y (2018) The toxicity of nanoparticles to human endothelial cells. In: Cellular and molecular toxicology of nanoparticles. Springer, Cham, pp 59–69

Chen L, Liu J, Zhang Y, Zhang G, Kang Y, Chen A, Feng X, Shao L (2018) The toxicity of silica nanoparticles to the immune system. Nanomedicine 13(15):1939–1962

Cui L, Wang X, Sun B, Xia T, Hu S (2019) Predictive Metabolomic signatures for safety assessment of metal oxide nanoparticles. ACS Nano 13(11):13065–13082

Ermolin MS, Fedotov PS, Malik NA, Karandashev VK (2018) Nanoparticles of volcanic ash as a carrier for toxic elements on the global scale. Chemosphere 200:16–22

Flühmann B, Ntai I, Borchard G, Simoens S, Mühlebach S (2019) Nanomedicines: the magic bullets reaching their target? Eur J Pharm Sci 128:73–80

Haque S, Whittaker MR, McIntosh MP, Pouton CW, Kaminskas LM (2016) Disposition and safety of inhaled biodegradable nanomedicines: opportunities and challenges. Nanomed Nanotechnol Biol Med 12(6):1703–1724

Hashempour S, Ghanbarzadeh S, Maibach HI, Ghorbani M, Hamishehkar H (2019) Skin toxicity of topically applied nanoparticles. Ther Deliv 10(6):383–396

Hou J, Wu Y, Li X, Wei B, Li S, Wang X (2018) Toxic effects of different types of zinc oxide nanoparticles on algae, plants, invertebrates, vertebrates and microorganisms. Chemosphere 193:852–860

Kim JH, Lee K, Jerng UM, Choi G (2019) Global comparison of stability testing parameters and testing methods for finished herbal products. Evid Based Complement Alternat Med 2019:7348929

Lameijer M, Binderup T, Van Leent MM, Senders ML, Fay F, Malkus J, Sanchez-Gaytan BL, Teunissen AJ, Karakatsanis N, Robson P, Zhou X (2018) Efficacy and safety assessment of a TRAF6-targeted nanoimmunotherapy in atherosclerotic mice and non-human primates. Nat Biomed Eng 2(5):279–292

Marques MR, Choo Q, Ashtikar M, Rocha TC, Bremer-Hoffmann S, Wacker MG (2019) Nanomedicines-tiny particles and big challenges. Adv Drug Deliv Rev 151:23–43

Modena MM, Rühle B, Burg TP, Wuttke S (2019) Nanoparticle characterization: what to measure? Adv Mater 31(32):1901556

Mohammed Abubakar B, Mohd Salleh F, Shamsir Omar MS, Wagiran A (2017) DNA barcoding and chromatography fingerprints for the authentication of botanicals in herbal medicinal products. Evid Based Complement Alternat Med 2017:1–28

Mukherjee PK, Harwansh RK, Bahadur S, Banerjee S, Kar A (2016) Evidence based validation of Indian traditional medicine–way forward. World J Tradit Chin Med 2:48–61

Musazzi UM, Marini V, Casiraghi A, Minghetti P (2017) Is the European regulatory framework sufficient to assure the safety of citizens using health products containing nanomaterials? Drug Discov Today 22(6):870–882

Paradise J (2019) Regulating nanomedicine at the Food and Drug Administration. AMA J Ethics 21 (4):347–355

Pferschy-Wenzig EM, Bauer R (2015) The relevance of pharmacognosy in pharmacological research on herbal medicinal products. Epilepsy Behav 52:344–362

Rodríguez-Ibarra C, Déciga-Alcaraz A, Ispanixtlahuatl-Meráz O, Medina-Reyes EI, Delgado-Buenrostro NL, Chirino YI (2020) International landscape of limits and recommendations for occupational exposure to engineered nanomaterials. Toxicol Lett 322:111–119

Sarwal A, Singh G, Singh K, Garg S (2018) Recent interventions for nanotechnology based drug products: insights into the regulatory aspects. Curr Pharm Des 24(43):5219–5228

Soares S, Sousa J, Pais A, Vitorino C (2018) Nanomedicine: principles, properties, and regulatory issues. Front Chem 6:360

Steinhoff B (2019) Quality of herbal medicinal products: state of the art of purity assessment. Phytomedicine 60:153003

Stuttgen V, Giffney HE, Anandan A, Alabdali A, Twarog C, Belhout SA, O′ Loughlin M, Podhorska L, Delaney C, Geoghegan N, Mc-Fadden J (2019) The UCD nanosafety workshop (03 December 2018): towards developing a consensus on safe handling of nanomaterials within the Irish university labs and beyond–a report. Nanotoxicology 13(6):717–732

Suk JS, Xu Q, Kim N, Hanes J, Ensign LM (2016) PEGylation as a strategy for improving nanoparticle-based drug and gene delivery. Adv Drug Deliv Rev 99:28–51

Surekha D, VS TR, Kumar A (2016) A review on role of quality control and quality assurance system in regulation of herbal drugs. Int J Res Phytochem Pharmacol 6(2):32–40

Thakkar S, Anklam E, Xu A, Ulberth F, Li J, Li B, Hugas M, Sarma N, Crerar S, Swift S, Hakamatsuka T (2020) Regulatory landscape of dietary supplements and herbal medicines from a global perspective. Regul Toxicol Pharmacol:114, 104647

Umezawa M, Onoda A, Takeda K (2017) Developmental toxicity of nanoparticles on the brain. Yakugaku Zasshi: J Pharm Soc Japan 137(1):73–78

Wacker MG, Proykova A, Santos GML (2016) Dealing with nanosafety around the globe—regulation vs. innovation. Int J Pharm 509(1–2):95–106

Watson CY, DeLoid GM, Pal A, Demokritou P (2016) Buoyant nanoparticles: implications for Nano-biointeractions in cellular studies. Small 12(23):3172–3180

GPSR Compliance

The European Union's (EU) General Product Safety Regulation (GPSR) is a set of rules that requires consumer products to be safe and our obligations to ensure this.

If you have any concerns about our products, you can contact us on ProductSafety@springernature.com

In case Publisher is established outside the EU, the EU authorized representative is:

Springer Nature Customer Service Center GmbH
Europaplatz 3
69115 Heidelberg, Germany

Batch number: 10739787

Printed by Printforce, the Netherlands